TABLE OF CONTENTS

A MANAGER'S GUIDE

TO

TECHNOLOGY FORECASTING

AND

STRATEGY ANALYSIS METHODS

by

Stephen M. Millett

and

Edward J. Honton

BATTELLE PRESS

Columbus • Richland

Library of Congress Cataloging-in-Publication Data

Millett, Stephen M., 1947–
   A manager's guide to technology forecasting and strategy analysis
methods/by Stephen M. Millett and Edward J. Honton.

      p.   cm.
   Includes bibliographical references.
   ISBN 0-935470-63-8 : $19.95
   1. Technological forecasting.   I. Honton, Edward J. 1955–
II. Title.
T174.M55   1991
658.4'0355—dc20   91-13690
                                                                        CIP

Printed in the United States of America.

Battelle Press
505 King Avenue
Columbus, Ohio 43201-2693
614-424-6393
Toll Free: 1-800-451-3543
Fax: 614-424-5263

LIST OF TABLES   Page

LIST OF FIGURES   Page

FOREWORD

Corporate managers, planners, and analysts ask many questions about technology forecasting and planning methods. Which forecasting method is most used today? What method is being used by which companies? How accurate are the methods? Unfortunately, there are no simple answers to these questions.

We think these questions reflect the confusion and misunderstanding that exist in many companies about technology forecasting and strategy analysis. Certainly forecasting and strategic business planning are a significant part of technology management. Yet the tools for managing technology as a corporate resource are not as well refined and understood as the tools for financial management or even those for personnel development.

Based on our experiences in workshops, conferences, projects, and consulting over the last decade, we have been sharing the following observations in response to the above questions:

1 No technology forecaster or analyst relies on just one, or even a few methods. Companies use many methods depending upon the subject, the goal, and available resources.

2 The variations in methods, techniques, and applications typically reflect the topic of analysis and the corporate culture more than the state of the art of the methods.

3 There appears to be a growing gap between decision-making

managers and information-producing analysts. Managers need answers, even if only approximations, quickly and in a manager-friendly manner as inputs to their decision-making process.

In this guide, we assess the current state of the art of 20 different forecasting methods, evaluate their advantages and disadvantages, suggest applications, and predict their future into the 1990s. Because we emphasize the merits and utility of the methods more than their mechanics, we clustered the methods into three broad categories:

1   **Trend Analyses**

2   **Expert Judgment**

3   **Multi-Option Analyses.**

We offer the following generalizations based upon this assessment:

1   **Too much emphasis has been placed on the accuracy of forecasts** and not enough on the educational and communication value of the forecasting process.

2   **Goals and purposes for the forecasts must be identified** before selecting the appropriate methods to achieve them. The very first question is "What is the question?" The second question is "What do you intend to do with the answer?" Then the methods are selected and applied accordingly.

3   **Methods should be used in combinations**. No one method can answer all questions. Trend analyses, expert judgment, and multi-option analyses can be combined according to the needs of the forecasting effort. How these methods are combined is very much dependent upon the skills of the analysts and managers and upon the corporate climate.

4   **Technology forecasting methods also need to grow** to incorporate applicable features of economic forecasting, political analysis, and market research. In the broadest sense, technologies, and especially their final products, are profoundly affected by nontechnical factors (such as economics, politics, public policy and regulation, social preferences, etc.).

5   **Technology forecasting and strategy analysis ideally should accomplish three goals:** first, provide a forecast of the future technological environment; second, suggest alternative technology strategies to managers; and third, evaluate these strategies to see which will produce the desired results.

6   **Nearly all the methods of technology forecasting and strategy analysis** that we examine **are also used for forecasting purposes outside technology.** The only methods that come close to being unique to technology are S-curves and patent trend analysis.

7   **We have observed recent growth in the number of specialists** within companies assigned to perform the specific job of technology forecasting.

We make some forecasts for the 1990s:

1   **Technology forecasting and strategy analysis methods will become better understood** and applied through their use in the corporate environment. The methods will become more sophisticated.

2   **More methods will be combined with each other.** All three categories of methods examined in this report show strong potential for improvement when used with other methods.

3   **The following methods will increase in popularity:**
  trend extrapolation
  time series estimation
  regression analysis
  historical analogies
  patent trend analysis
  scientific literature analysis
  analysis of user created databases
  interviews
  questionnaires
  idea generation
  nominal group technique
  scenarios
  simulations.

4   **The following methods will decline in popularity:**
econometrics
systems dynamics
S-curves
input-output matrices
Delphi method
paths and trees
portfolio analysis.

5   **We are aware of only one new forecasting and strategy analysis method** that will emerge as newly popular in the 1990s, namely the creation of a user's own database for electronic analysis. There may be other newcomers of which we are not aware. However, such new tools are highly unlikely to replace the more then twenty methods considered in this guide within the next decade.

We conclude with several recommendations for managers. In general we believe managers need to

1   **Clarify their needs to analysts** so that they can be more responsive to managers' needs for decision making.

2   **Prepare a written inventory of existing tools and data-bases** presently used by the company. Analyze the inventory to ascertain whether a transfer of existing knowledge, software, or data from one working group to another would be beneficial.

3   **Examine the trends for the 1990s** in forecasting tools **and explore what these trends mean** for the company. Does changing software or data availability suggest any changes the company should make in the use of tools?

4   Request that analysts **evaluate any specific tools or data-bases** mentioned in this report **that promise to improve the company's forecasting ability** or provide a better understanding of technology problems.

5   **Implement the use of multiple methods in various combinations** for better technology forecasting. While there are a great number of possible combinations of methods, one array seems to be particularly attractive:

A   Expert judgment (particularly interviews and surveys) to frame the right question for the forecasting study;

B   Expert judgment (particularly idea generation and the nominal group technique, or variations thereof) to identify issues, factors, trends, variables, etc. to be included in the scope of the forecasting study;

C   Trend analyses (particularly trend extrapolation, time series, and patent trend analysis) to understand thoroughly the past, present, and most feasible future of each factor in the scope of the forecast;

D   Multi-option analysis (particularly scenarios and simulations) to integrate the trends and to generate alternative, including normative, views of the future;

E   Expert judgment (particularly idea generation and the nominal group technique) to draw business implications and strategic options from the forecasts; and

F   Trend analyses (particularly trend extrapolation and time series) combined with other forecasting methods (especially econometrics and financial projections) to do detailed, microscopic analysis for planning purposes.

6   **Expand their concept of technology forecasting toward** the broader concept of **product forecasting**, which includes business environment and corporate concerns as well as technological performance.

7   Incrementally **integrate changes required by the new technology forecasting methods** into the office. This includes acquiring appropriate computer hardware and software, ensuring adequate access to the proliferating number of electronic databases, and training staff to use both new forecasting tools and combinations of tools.

C H A P T E R 1

INTRODUCTION

Many companies now realize that technologies, especially intellectual properties in the forms of copyrights, patents, and licenses, are corporate assets that must be managed like money, facilities, and people. A significant part of technology management is forecasting and strategic business planning. Yet the tools for technology planning are not well refined or understood. Because of the number and type of questions asked about technology forecasting, we prepared this guide to clarify certain issues about forecasting and analytical methods so that managers can better appreciate and use the excellent work of many skilled analysts.

The information contained in this guide came from many sources. We collected and reviewed articles, papers, and books on the topics of forecasting, business strategy, and R&D management (a few of which we authored ourselves). We also drew upon dozens of our forecasting experiences for corporate clients in the United States, Europe, and Japan. We carefully used the confidential remarks of strategic planners and managers from hundreds of companies with whom we have formed friendships from our workshops and field visits. We have enjoyed direct contact with business people in the United States, Canada, Mexico, the United Kingdom, Ireland, the Netherlands, France, Portugal, Italy, Switzerland, Germany, Norway, Finland, Japan, and Korea. We have provided technical consultation to a wide variety of companies, including oil, electric, and gas; automobile; telecommunication; aerospace and defense; postal service; consumer products; medical equipment; and health care, as well as governmental development agencies. Ultimately, this guide is the culmination of our ten years of Battelle work in the area of forecasting, strategy-making, and R&D management.

This guide is organized into three principal chapters: (1) **Trend Analyses** (including trend extrapolation, time series, regression analysis, systems dynamics, S-curves, analogies, and patent trend analysis); (2) **Expert Judgment** (including interviews, surveys, Delphi method, idea generation, and the nominal group technique); and, (3) **Multi-Option Analyses** (scenarios, simulations, paths and trees, and portfolio analysis). A conclusion follows with generalizations and our own expert judgment about developments toward the year 2000.

Before beginning our review of the methods, we need to discuss briefly two themes common to all of the above forecasting and strategy analysis methods. First, we need to define "technology forecasting" because there is confusion about what it is and is not, as compared with economic forecasting or weather forecasting. Second, we need to establish the context for technology forecasting by examining the stages of technological development common to many corporations.

Professor Martino in his popular textbook, *Technology Forecasting for Decision-Making,* defines technology forecasting as "prediction of the future characteristics of useful machines, procedures, or techniques." The accent is upon practical applications, not on scientific knowledge. It is also upon the characteristics, or parameters, of practical performance, not upon the actual appearance of the technology. Technology forecasting is not the same as product or market forecasting, although they are often confused. Managers typically think like customers: what is the product, what does it do, and what is its value? They often expect technology forecasts to address these questions, and they are disappointed when the forecasts do not. On the other hand, engineers often confuse technology with products and make claims that are difficult to support with the tools of technology forecasting alone.

William Ascher in his acclaimed book, *Forecasting. An Appraisal for Policy-Makers and Planners,* defines technology forecasting as the effort "to project technological capabilities and to predict the invention and spread of technological innovation..." He agrees with Martino that the subject is technological capabilities, but he differs on other aspects of the definition. Indeed, Ascher goes further than either Martino or we would wish to go in defining technology forecasting.

What, after all, is meant by "forecasting"? In a narrow sense, the word means prediction (which Martino uses); and quantitative analysts think of prediction as a specific number. In a broad sense, however, forecasting is the expectation and estimation of future conditions typically bounded by ranges or described in words. Shell International, for example, likes to call what it does "foretelling" rather than "forecasting," precisely because Shell has yielded the broad definition of the term "forecasting" to the narrow definition of quantitative analysts. We still hold to the broad definition.

Therefore, we define technology forecasting as the process and result of thinking about the future, whether expressed in numbers or in words, of capabilities and applications of machines, physical processes, and applied science. Technology forecasting certainly involves more than projection, and it does not have to be so specific as a prediction. It does not have to describe the actual invention and spread of technology; but if it can, it would make the forecast more comprehensive and useful to managers. One of the challenges of the 1990s is to stretch forecasting methods further beyond the domain of technology and to integrate them with other types of forecasting methods, such as those used in economics, politics, and meteorology.

Indeed, in the corporate context that we have in mind, "product forecasting" may be more descriptive of what we are really attempting than "technology forecasting." This new term opens up horizons for integrating technologies with nontechnical considerations in a comprehensive package that would be more useful to managers.

We completely agree with Martino that technology forecasts should include four elements:

1   a specified time period (such as five or ten years from now or dates such as 1992, 2000, 2010, etc.)

2   the technology domain (mechanics, approach, and topology in general terms, at least)

3   performance characteristics of the technology, especially the parameters (defined and described, if not specified in numbers)

4   likelihood or probability of occurrence by the specified time.

We further agree with Martino that accuracy is not the sole, or even major, emphasis of technology forecasting. Our experience has been consistent with his in that the value of a forecast is measured by its usefulness to a manager faced with a decision. Both managers and analysts have overstressed the importance of accuracy by believing and acting upon forecasted numbers. A major short-coming of too many forecasting tools is that they produce answers without explanations, and managers compound the problem by accepting numbers without further discussion. Astute managers question vigorously the methods, results, and business implications of forecasts. As forecasters and analysts ourselves, we want greater accuracy; but we realize that accuracy is not the only benefit of forecasts, and that inaccurate forecasts can produce positive benefits. Beyond accuracy, the timeliness and packaging of the forecast are extremely important to managers.

Forecasting is usually performed along with planning. Unfortunately, much corporate planning is as stilted as forecasting. We even hate to use the word "planning" because too many business people associate it with boring meetings, lengthy reports, and soon-forgotten distractions to the day-to-day business of making money. Planning was never meant to become a substitute for thinking, but too often in practice it has. Today, corporations need more "strategic thinking" and less so-called "strategic planning." We prefer to use the term "technology forecasting and strategy analysis methods" to mean ways of thinking about future technological progress and new product development with respect to both external business environments and internal corporate cultures. The goal is to evaluate and select the corporate strategies for actions that are most likely to produce desired results. Ultimately forecasters and analysts present their conclusions and managers take action.

Several authors have presented models of the R&D process in terms of discreet stages of technological development. Because an overview of the innovation process does provide a practical context for technology forecasting methods, we would like to present our own model of the process:

1   **Early theorizing and conceptualization.** This is the "light bulb" phase when someone comes up with an intuitive bright idea. The idea at this stage has little form and substance, but much enthusiasm.

2   **Exploratory research.** Will the idea work? What form will it take? At this stage the researchers need to achieve some

critical parameters. Scientific methods, plus a lot of trial and error, are used to manifest the early idea, which is often modified in the process.

3 **Component development.** If the second stage produces promising results, the next stage is to develop and fabricate the parts that will make the emerging technology work.

4 **Prototype development.** If the third stage succeeds in its goals, the components will be assembled into the prototype product.

5 **Testing of prototype.** This may involve both physical testing of the assembled technology and market testing for possible consumer demand.

6 **Initial manufacturing and marketing of the product.**

7 **Consumer acceptance or rejection.**

8 **Product modification and improvements.**

9 **Product maturity and decline.**

Technology forecasting and strategy analysis methods can be employed at virtually any one of these stages, but they may have different applications. In the first three stages, the forecast is most likely to conform to the strict definition of technology forecasting. At the later stages, forecasts and analyses more likely take the shape of product, market, and economic forecasts. By Stage 7 we are strictly faced with product forecasts rather than technology forecasts. The manager needs to appreciate the stage of technological innovation in which the forecast is conducted. The analyst also needs to understand this context for the forecast and appreciate the decision needs of the manager who will be using the forecast.

One challenge for analysts and managers in the 1990s will be to explore innovative ways of combining and applying the methods of technology forecasting. Another challenge will be to integrate technology forecasting with market research methods. Market research methods are not covered in this study, although we believe that exploration in this direction by technologists is indeed merited to achieve the full potential of "product forecasting." Our unrefined impression is that generally the methods of market research are similar to those of technology forecasting, especially the variations on expert judgment techniques, including surveys, interviews, and focus groups. Sales projections also strike us as being essentially trend projections combined with expert judgment. In market research, more than in technology forecasting, the "experts" in "expert judgment" are the target customers themselves, not the R&D technologists. Technology forecasters and market researchers have a lot to learn from each other.

We are very pleased to see a new appreciation for expert judgment methods. The July–September 1990 special issue of the *Journal of Forecasting* focused on expert judgment methods. The editors acknowledged that "Together, these papers question the assumption that statistically based forecasting is superior to forecasts made by the exercise of human judgment...Five years ago we could not have foreseen that the role of judgment in forecasting would play such a significant role as it does now, but some things are difficult to forecast!" Indeed, on some matters, analysts have more difficulty preparing forecasts than managers have in using them, because managers have long appreciated the importance of expert judgment in making decisions about today's investments and tomorrow's returns.

2

TREND ANALYSES

Trend analyses have been and still are the most popular technique used for technology forecasting. There are good reasons for this popularity. Trend analyses are better than guesses, yet can still be relatively inexpensive to perform. The basic idea is simple— collect relevant historical data and then graph it to project a trend.

Sometimes a trend projection is portrayed graphically, sometimes numerically. Sometimes the projection is made using only a simple one-variable trend, sometimes with many variables operating in complex relationships with each other. Although trend analyses are done in many different ways, they all have some common assumptions and features, namely:

1   The future is a continuation of the recent past and, because human behavior follows natural laws that can be expressed quantitatively, as in physics and chemistry

2   There is one future and it is predictable if you understand the underlying laws as shown in the trend data.

Before reviewing any particular method, it is important to note that the assumption that the future is a continuation of the past brings into the forecast the same risk of inaccuracy as does a judgment of which variables to use as the basis for a multi-option forecast. In other words, in trend analyses the statistics themselves are often allowed to establish the choice of important variables, simply because future relationships are held to be equally as valid as past relationships. Either way a judgment is being made. The use of numbers and statistics should not hide this assumption. One point to remember is that all forecasts are conditional in that they are predicated on a set of parameters holding true (either past data or assumptions).

The second assumption, that a single knowable future exists, leads to a statistical process in which a forecast is based on projecting historical data. In doing this, it is desirable to select variables that respond to cause and effect relationships. Normally it is not a good idea to forecast based strictly on the existence of a statistical relationship, but rather the forecast should be based on some meaningful theory of cause and effect.

Trend analyses can be dangerous when used without a theory of causation. Bizarre forecasts can result. For example, it is a known statistical fact that U.S. stock market performance can be predicted based on which football conference wins the Super Bowl. The *Wall Street Journal* recently found another statistical relationship between stock market performance and which team wins the Super Bowl. The new relationship correlates the team name that comes first alphabetically with stock market performance over the previous six weeks. The point of these two examples is that there is no meaningful theory as to why and how the variables are related.

A proper example is that the level of R&D expenditures in a country appears to be related to the growth rate of that country's economy, its savings rate, and its interest rates. A regression analysis of R&D expenditures could be performed on these three variables to measure the actual effect of each variable on R&D expenditures.

The concept of using trend analyses to predict the future is actually quite old. The earliest human civilizations examined evidence around them to decide when to plant crops, when to harvest, etc. In modern times, the increase in use of statistics has been directly related to the increase in machine computing power. During the 1960s, in particular, computers fueled a large increase in use of complex econometric models over more simple trend extrapolation. It is anticipated that increases in computer power will continue to affect the type of trend analyses performed.

As indicated above, many trend analyses tools are used today for various technology forecasting purposes. In the sections that follow we review ten of these tools:

- Trend Extrapolation
- Time Series Estimation
- Regression Analysis
- Econometrics
- Systems Dynamics
- S-curves
- Historical Analogies
- Input-Output Matrices
- Patent Trend Analysis
- Scientific Literature Analysis.

In an additional section, we offer a few thoughts about a new way of forecasting using published data sources that is likely to become popular during the 1990s.

Trend Extrapolation

Trend extrapolation is the simplest form of technology forecasting. The underlying concept is that the future will be a simple extrapolation of the past. Information is collected about a variable over time, and then extrapolated to determine where that variable will head in the future. Trends are often portrayed graphically.

For example, examine Figure 1 for a trend extrapolation of the efficiency of man-made white light. Efficiency is presented on a logarithmic scale, rather than a linear scale, because efficiency has actually been accelerating in growth in recent decades. The line drawn through the points suggests that, by 1990, white light should reach about 500 lumens per watt.

While this may be achievable, it is unlikely that this same trend extrapolation will hold into succeeding decades, thus we have drawn an arrow that indicates that the actual data may be much lower than the theoretical extrapolation. This illustrates a major problem with trend extrapolation—often the simple line forecasts are only believable over a short time frame of one or two years—and expert judgment may suggest the extrapolation will not hold over a 20-year period.

Trend extrapolation has been used to forecast technological capabilities, level of product sales, and the length of time it will take to develop a new technology, among many other events. In practice, almost every company in every industry collects historical trend information on important variables as a starting point. This database of past and current technological capabilities is quite valuable for

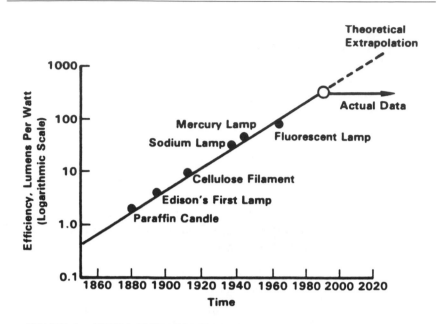

FIGURE 1.   TREND EXTRAPOLATION OF EFFICIENCY OF WHITE LIGHT

understanding the near-term future. Simple graphs allow a forecast line to be drawn into the future. This is often better than a guess and is, in fact, used by many companies for making a better "educated guess."

Trend extrapolation is relatively inexpensive and can be performed quickly, if the necessary data are available. Personal computer software products that extrapolate data abound today. For example, the highly popular Lotus 1-2-3™ program can plot an extrapolation in a matter of seconds, as can numerous other software packages. The level of training required to obtain and interpret the results is really very minimal. Overall, trend extrapolation is often the first step in a technology forecast, if for no other reason than to provide a simple first approximation.

The primary disadvantage of trend extrapolation is that the line drawn is often wildly inaccurate. Circumstances affecting the forecasted variable can change remarkably over time—and trend extrapolation cannot account for a change due to any cause. The longer the time period of the forecast, the less likely the forecast will bear any resemblance to actual events. Generally, trend extrapolations are good only for a few quarters up to one year. It is a mistake to make middle- to long-term forecasts based solely on an extrapolation.

Our own extrapolation is that trend extrapolation will continue to serve a short-term forecasting role during the next decade. With today's computers and software, it is likely that more managers will perform their own calculations without help from analysts. This trend will be bolstered as computer networks come into the office during the decade, providing easy access to ever increasing amounts of company, industry, national, and global data.

Time Series Estimation

Time series estimation takes trend extrapolation one step further. An effort is made to distinguish statistically the systematic from the random variation of a trend over time. The objective is to predict the future by projecting the knowable systematic variations. In other words, time is a surrogate measure for all the causes affecting the variable being forecast. Statistics do the rest.

Several variations might be measured including seasonal effects, periodic cycles, trends, known one-time special effects, and random effects that cannot be measured systematically. Seasonal effects are most often measured either quarterly or monthly. Periodic cycles vary in length from months to many years, and may be used to measure the innovative cycle of applying new basic technologies to existing product areas. Trends are usually measured exactly as in trend extrapolation—an effort is made to detect either an increasing or decreasing linear, logarithmic, or exponential line. A one-time special effect is often the result of a technological breakthrough, such as the introduction of the transistor and its subsequent application as part of many new technologies.

A large number of statistical techniques have been developed to extract such systematic variation. They include moving averages, exponential smoothing, Box-Jenkins, decomposition, and multivariate regression analysis. These techniques are taught in most college statistics courses, and the reader is referred to a college textbook for the mechanics of these methods.

As an example of time series estimation, the daily sales of newsprint in France are graphed for several years in Figure 2. There is a basic upward trend over time, along with a large cycle in the data, and some shorter period cycles. The object of time series estimation is to fit a curve to this data that gives the best forecast beyond 1980.

A time series statistical fit usually is more accurate than a simple extrapolation. For example, a trend line drawn for the data

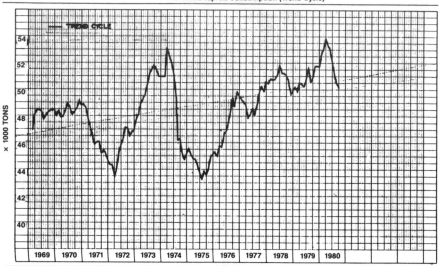

Source: Makridakis, 1982, p. 160. Reprinted with permission.

FIGURE 2.   TIME SERIES ESTIMATION

in Figure 2 will not provide as accurate a forecast as a curve consisting of a basic trend line modified by long- and short-term cycles. In fact, in a recent review of economic model forecasting accuracy, Victor Zarnowitz (1986) states, "More attention to data and techniques that are sensitive to business cycle movements and turning points could help improve their record."

Time series estimation is also a slightly more complex calculation than a trend extrapolation because a Lotus 1-2-3 graph no longer suffices. Instead, the various available time series statistical routines must be each run on the data to see which provides the best statistical fit. The best method is simply the method that gives the best fit. There is no a priori reason to know which technique to use in any given circumstance. The goal is to explain as much of the variation in the time series as possible while minimizing random variation. Also, the best fit equation produced is assumed to provide the best forecast, which empirically is not always true.

To make the process far less time consuming and expensive, several computer software packages have been developed. Popular software includes the Statistical Analysis System (SAS), the Statistical Package for the Social Sciences (SPSS), the Time Series

Program (TSP), and the Econometric Software Package (ESP)—all are available for both mainframe and personal computers. Using such a program, one can explore time series variation in a set of data with about an hour of work. The final equation or graph produced can then be readily used to forecast. The only real problem with this practice is that many analysts today are using such data crunching techniques without understanding how to interpret the statistical results and, in fact, without knowing whether the application they are forecasting follows the assumptions of a particular statistical procedure. It is wise to spend a few days learning about the Box-Jenkins, exponential smoothing, decomposition, or moving averages techniques before applying them to data.

Popular uses of time series estimations include projecting technology performance characteristics and product sales quarter by quarter. Most companies in most industries use this technique in some part of their business.

The primary disadvantage of the method, as with trend extrapolation, is that causation is not accounted for in explaining the variation. Thus, a variable can move substantially away from its existing trend, and time series estimation will not account for the change. It is, therefore, inappropriate to use time series for middle- to long-term forecasts. In addition, time series requires more effort than trend extrapolation (about an hour instead of five minutes) because it is slightly more complex.

As with trend extrapolation, time series estimation will likely increase in use, as a first forecasting step, over the next decade. The rapidly increasing availability of time series data in electronic form will enhance this growth.

Regression Analysis

Regression analysis is a generalized form of time series forecasting; however, instead of forecasting a variable over time, a regression is used to forecast a variable as a function of one or more other explanatory variables that may or may not include time. For example, to predict the life expectancy of an automobile tire, tire mileage may be assumed to be a function of explanatory variables such as the hardness of the rubber, number of steel belts, amount of other materials, thickness of the tread, shape of the tread, weight of the automobile on which it is used, and numerous other factors related to where the auto is driven. Data are collected on each of these variables and coefficients are statistically estimated that provide a best fit curve of the useful life of the tire to the explanatory variables. In addition, numerous statistics indicating the precision and reliability of the fit can be calculated. (Analysts love regressions.) Then, given forecasts for each explanatory variable, and these coefficients, a value for tire life expectancy can be forecast as well.

Regression analysis is commonly used, both for time series and for non-time series forecasts, and is easy to perform on today's computers. Numerous programs (including SAS, SPSS, TSP, and ESP) make regression analysis simple and inexpensive, provided data exist for each variable in a number of different cases. An analyst can normally construct a good regression model in an afternoon of work.

The primary advantage of regression analysis over time series projection is that a regression is calculated assuming a degree of causation. The forecasted variable changes as a consequence of changes in the explanatory variables. For example, if automobiles generally become lighter, tire life will be extended. Such technology performance estimations are the primary use of regression analysis.

A good forecast depends both on selecting the correct explanatory variables and on knowing the expected value of the explanatory variables. Hence, regression analysis is not suitable for forecasts in which the causes for a change in a forecasted variable or in which the future values for an explanatory variable are unknown. Thus, regressions are not very useful for estimating when a new product will be introduced into the market or for predicting what new manufacturing processes will exist in five years.

Overall, regressions are more complex than time series, and often produce better forecasts, given the availability of better information and additional analytic effort.

As with trend extrapolation and time series estimation, we see regression analysis being used at least as much over the next decade as it is presently. And it will largely be used for many of the same technology forecasting purposes as time series estimation such as estimating technology performance characteristics and product sales. Regression analysis may even become slightly more popular because computer programs are easier to use, personal computers are common in the office, and there is growing access to electronic data bases. In addition, as time goes by, more and more of the causal relationships between variables related to technology performance will be discovered through academic studies, which will help an analyst who is constructing a real world regression arrive at a meaningful forecast sooner and less expensively.

Econometrics

An even more sophisticated method for technology forecasting using underlying trends is econometrics. This technique bundles several regressions into a model of interrelationships.

As a simple example, suppose we wish to know what portable personal computers are likely to weigh next year or in three years. The main equation that answers this may forecast the weight of personal computers as a function of battery weight, availability of low-power consuming flat plate displays, number of disk drives, and the amount of hard disk storage available in the machine. In turn, battery weight may be forecast at least partially as a function both of the rate of technical advance in sodium-sulfur batteries and the increase in the temperature at which materials become superconductive. In a third equation, the availability of flat plate displays that consume relatively little power may also be directly related to advances in sodium-sulfur or other advanced technology batteries. Finally, the rate of development of new superconductive materials may be forecast partly as a function of the degree of inexpensive and readily available personal computers with large memory storage capabilities. This circular set of equations is interrelated in such a way that the overall central forecast of portable personal computer weight over the next few years must come from the simultaneous solution of four equations.

The first step in econometrics is to decide which equations to estimate and how the equations will interrelate. Secondly, the equations are calibrated using historical data to establish the nature of the causal relationships, often measured using regression analysis on each equation. Finally, technology forecasts are made using the calibrated model. In sum, an econometric model is a causal model that uses many variables within a set of multivariate regressions.

Econometrics are appropriate when (1) causal relationships can be identified, (2) there has been a large change in the causal variables over time, and (3) it is possible to predict the direction of change in causal variables. This last requirement exists because the forecast does not depend simply on changes occurring over time. Rather, as in the simpler regression analysis, there are a number of causal variables that help determine the future of the predicted item. In the personal computer weight example, it is necessary to predict the number of disk drives to be used in each computer, the amount of hard disk storage available, and the rate of advance in high technology batteries before the overall model can be used to develop the forecast.

Clearly, building such a model is more expensive than estimating single variable forecasts as in trend extrapolation, time series estimation, or regression analysis. As an advantage, however, the very nature of the interrelating equations often causes a properly calibrated model to forecast results that cycle up and down over time—that is, the forecast is prevented from growing or declining indefinitely over time so that the result is bounded within a realistic range. For example, an econometric model of efficiency of man-made white light would not increase enormously in the 21st century as it does with the simple trend extrapolation.

While we can conceive of econometric models that assist in R&D planning, in forecasting the introduction of new manufacturing processes, or in forecasting new product developments, we are not aware of many companies that use econometric models for these purposes. A few planners have told us they use econometrics to obtain a better handle on product sales forecasts by estimating several equations covering their customer demand, their own manufacturing supply, and technology factors within a simple model framework.

One drawback to econometric models is that they are often fairly complex and can be relatively expensive to construct. With a good software package, such as ESP or TSP, it usually will take several weeks of work by an analyst, together with a technologist, to prepare the model. Even with this investment of time and research, the final result may not greatly increase either accuracy or understanding.

An example outside the realm of technology forecasting illustrates this point. An econometric model with several equations was constructed to predict who would win the U.S. presidential election in 1988. Equations covered the demographics of the population, regional economic performance, the money supply, the deficits, and growth in gross national product, among other variables. All these equations were interrelated. The model produced a fairly accurate anwer. However, a simple regression model of one equation with four variables can also be used to predict the outcome of the election— the rate of growth in the economy over the last two years, the inflation rate over the last two years, the political party holding the White House during the last term of office, and whether the U.S. had been involved in a war during the last four years—in short, a peace and prosperity model. The simple regression produces a forecast as accurate as the more complex econometric model. Both are within two percentage points of the actual vote for all elections in the 20th century.

Similar results have been found when forecasting national economic performance. Econometric models with ten or so equations are often nearly as accurate as models with 1000 equations, and are much easier to comprehend and use. The message is clear: increased complexity does not necessarily lead to better understanding or accuracy.

One final caution about econometrics applies just as it does in regression analysis; namely, the forecast will likely not be accurate over the long term because either (1) structural changes in the economy are not measured by the variables and equations in the model or (2) values for the explanatory variables are not properly predicted over the long term.

Our forecast is that econometrics will decline in popularity over the next decade. They were much in vogue during the 1960s and 1970s when newly developed computers first made them realistic. However, during the last ten years many people have recognized their limitations in terms of their cost compared to the accuracy and understanding they provide. During the 1990s they are not likely to regain their former popularity, and we believe their use will decline somewhat in the corporate world.

Systems Dynamics

Systems dynamics solutions are based on cybernetic theories and systems analysis techniques developed at MIT, largely by Jay Forrester (see Forrester 1961, 1968, and 1971). The theory behind systems dynamics is that most events in life occur cyclically and should be modeled accordingly. Hence, systems dynamics models factors and influencing variables that interrelate through a set of feedback loops. The simultaneous solution of all feedback loops often produces forecasts that are bounded and fluctuate up and down in a regular pattern over time. The result is similar to econometrics, but systems dynamics solutions often are more regularly cyclical, even approximating a sine wave in shape.

An example of a systems dynamics model is given in Figure 3. The rectangles represent the central factors to be forecast. The circles represent influencing variables that partly interconnect the factors. The arrows show the feedback loops between the central factors and influencing variables. The arrow heads indicate the direction of the feedback.

In practice, the solution of a systems dynamics model is accomplished numerically. Each of the factors, influencing variables,

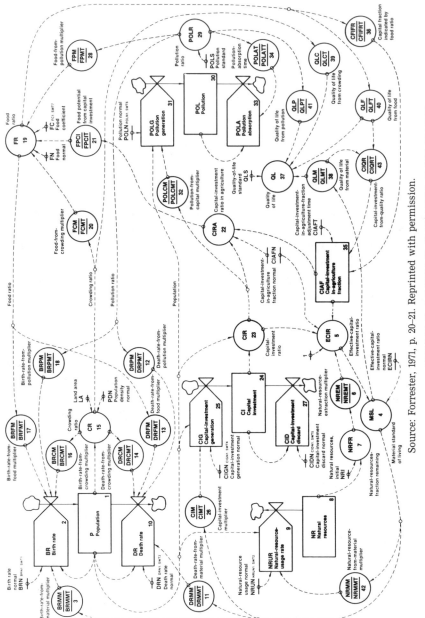

Source: Forrester, 1971, p. 20–21. Reprinted with permission.

FIGURE 3. SYSTEMS DYNAMICS FLOW CHART

and feedbacks is numerically estimated. Then the model is calibrated using historical data to make certain the selected numbers make sense. Finally the calibrated model is solved simultaneously over time into the future to make the forecast.

A well-known study, *The Limits to Growth,* was prepared in 1970 using systems dynamics on 66 key world factors (see Forrester, 1971, and Meadows, 1985, for a discussion of this study). This was the first computer global population growth model ever prepared. It showed, using feedback loops, that without deliberate action to slow world population growth, the most likely scenario would be "overshoot and collapse," resulting in a destruction of the resource base and a major decline in population. The baseline results are shown in Figure 4 This figure also illustrates the sine curve cyclical nature of systems dynamics results.

Meadows points out that such global models are often misunderstood.

> Global models are not meant to predict, do not include every possible aspect of the world, and do not support either pure optimism or pure pessimism about the future. They represent mathematically assumptions about the interrelationships among global concerns such as population, industrial output, natural resources, and pollution. Global modelers investigate what might happen if policies continue along present lines, or if specific changes are instituted.

These points are useful to remember. We too believe that a systems dynamics model is useful for thinking about how factors interrelate, and possibly for strategy analysis, but is not appropriate as a forecasting tool.

Systems dynamics models are rarely used in technology forecasting today. As hinted above, the major purpose of systems dynamics is to get managers and technical experts to identify feedback loops. Their specification, and the subsequent drawing of a figure indicating how the variables interrelate, is usually the end product. This thought process can be applied to a wide variety of forecasting problems, including establishing the steps necessary to innovate within a particular industry or identifying the variables likely to affect the development of a new process or product and determining how these variables interrelate.

Today, companies seldom actually specify numerical values to solve a systems dynamics model. Rather, this technique is most often

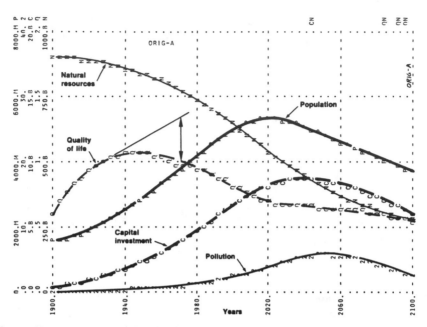

Source: Forrester, 1971, p. 70. Reprinted with permission.

FIGURE 4.   SYSTEMS DYNAMICS SAMPLE SOLUTION

used to provide input into either a trend-impact or cross-impact model. There are three reasons for the limited use of systems dynamics in industry. First, it is neither widely known nor has it been widely proven as a valid technique for technology forecasting. Second it is complex, and the final graphic does not seem to reduce the complexity. In this respect, it is quite difficult to communicate the results to someone not involved in creating the model. Third, many specific problems such as estimating technological performance capabilities, or modeling the steps to innovate a new design or product, are actually more linear in nature than cyclical. Thus a model that proceeds from step to step linearly may provide a better understanding of reality in these situations.

Another disadvantage of systems dynamics is that no generalized software exists that permits creating a model from scratch in a new technological area. Each model must be created individually. This makes the process both time and computer intensive and, therefore, expensive. Given this cost, together with the difficulty of applying the results to real business situations, it is hardly surprising that the technique is not well known or used.

We believe these problems will persist throughout the 1990s. Even if generalized software is created to assist in systems dynamics modeling, the problems to which it can be applied are still limited to those where a cyclical nature to the forecast is anticipated. If you are dealing with a problem that has many interrelating variables, it is better to use either econometric modeling or cross-impact analysis. Software tools to assist with these methods already exist.

In each of the five methods discussed so far, historical data is used in an ever more complex way. Trend extrapolation merely indicates where a single line is headed over time. Time series estimation measures trends and cycles in the time series data to give a more accurate forecast. Regression analysis takes this the next step by making causal forecasts involving several variables that are not necessarily time dependent. Econometrics goes another step by combining several regression equations in a complex set of causal relationships. Finally, systems dynamics assumes that a large number of feedback loops exist between each forecasted equation, making it a specialized cyclical form of an econometric model.

The cost in time and money of implementing each of these methods increases with their complexity; and in practice, their use declines with complexity.

In the next three sections, specialized forms of trend analyses known as S-curves, historical analogies, and input-output matrices, are presented.

S-Curves

S-curve analysis recognizes that a technology introduced into the marketplace has a characteristic life cycle. In 1962 Fisher and Pry first demonstrated that two technologies competing for market share follow an S-shaped curve in market sales (see Fisher and Pry, 1971).

The pattern represented in Figure 5 illustrates the theory of S-curves by showing that the market share of a newly introduced product grows very slowly at first. During this first stage of its life cycle the product is introduced into the market, its capabilities become known, and it slowly gains sales momentum. During the second stage, growth in market share occurs rapidly and the slope is much steeper. This happens as the product matures, marketing plans succeed, and repeat sales or word-of-mouth sales are made. Finally, during the third stage of the life cycle, the market share tapers off as the product saturates the marketplace, competing

LIFE CYCLES

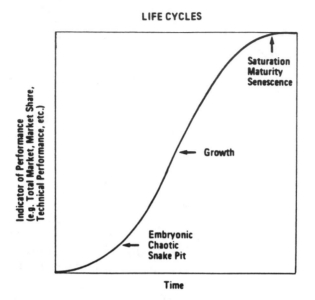

Source: Lee and Nakicenovic, 1988, p. 414. Reprinted with permission.

FIGURE 5.   GENERALIZED S-CURVE OF TECHNOLOGY LIFE CYCLE

products limit growth, and the technology upon which the product is based becomes obsolete. Eventually, market share stabilizes at an asymptote. (Beyond S-curve analysis, the actual level of sales will likely fall from the asymptote as the product itself becomes obsolete and is replaced by better technologies and products. That is, the market share may remain steady, but level of sales will fall as the overall market demand dwindles.)

The S-shaped life cycle concept has been shown both to work and not to work in many competing technology circumstances since the original Fisher-Pry work. While the concept is valuable, it cannot be taken too literally. Forecasts based on S-curve principles have helped some companies compete, and ignoring S-curve principles have hurt other companies. For example, some argue that during the 1970s the electric utility industry ignored the maturation of electricity as an energy form competing with natural gas, among other forms; and, as a result, the electric industry today is plagued with excesses of capacity and a consequent capital squeeze.

The S-curve pattern may be applied to not only market penetration of new products but also to the rate of dissemination of a new technology's use within products, or to the performance characteristics of a technology.

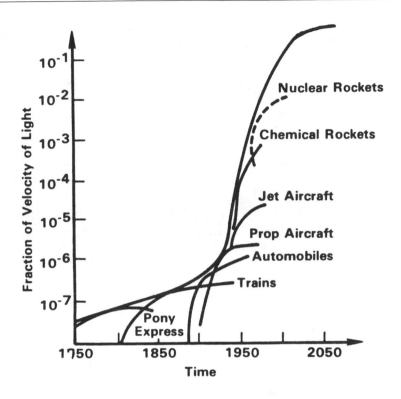

FIGURE 6. S-CURVE OF TRANSPORTATION SPEED

The example in Figure 6 is actually a family or envelope of S-curves. Individual technologies are graphed showing their perform-ance over time. The first commercial trains were introduced early in the 19th century, rapidly gaining in speed capability through 1860, and then tapering off to little additional speed capability after 1900. This same S-shaped pattern appears for several transportation modes. The set of curves, represented by the bold line across the top of all the individual curves, gives the S-curve for transportation speed. In addition, the S-curve pattern indicates that a new tech-nology, such as nuclear rockets, may provide the best forecast for maximum transport speed during the coming decades.

In this example, a natural physical asymptote for the curve is the speed of light. For many S-curves in technology forecasting, such a natural barrier exists. For example, market share cannot exceed 100 percent. As another example, for all practical purposes, the temperature at which materials become superconducting need not exceed room temperature, as most practical applications would then not require any cooling. (In fact, a recent theory of how supercon-duction occurs suggests a real asymptote of about −140 degrees F for such materials.)

The key to forecasting with S-curves lies in having previous experience with a similar technology. For example, one way to forecast the expected speed of new supersonic transport (SST) airplanes based on methanol fuel use would be to look at the history of jet engine aircraft subsequent to Whittle's invention of the jet engine in 1936. In the case of the jet engine, the first real use on an airplane was in the early 1940s; and in subsequent decades the power of the engine, combined with improved airplane design, allowed increasing speed records. Perhaps the SST technology will follow a similar curve of development, both in terms of years to first commercial use and in terms of growth in speed performance.

This example points out the drawback to using S-curves—uncertainty about which previous technology to compare with a new entrant. Much effort has been expended to prepare representative gamma coefficients that allow the full curve to be drawn based on the historical trend to date. However, we have not found any companies that regularly use this method as a forecasting tool, primarily because of uncertainty as to the appropriate gamma coefficient.

Despite these difficulties, recent research still shows that S-curves are useful. Thomas Lee and Nebojsa Nakicenovic (1988) argue that "S-curves may not be reliable forecasting tools, but they can be very useful for contingency planning." In addition, when more than one attribute of an examined technology appears to follow the S-curve pattern, they show that "a single S-curve is not as useful, for planning purposes, as a set of S-curves dealing with different aspects of the same technology and market." In Lee and Nakicen-ovic's view, the two major uses to which S-curves can be applied today are investment and R&D decisions and contingency planning.

The most popular use of S-curves is as a reminder that the current growth rates in market penetration, the diffusion of technology, or technological capability will not and cannot continue forever. By graphing the historical data over time, as in a trend extrapolation, and remembering that some asymptote does exist, it is possible to add to the educational content of an educated guess. Like systems dynamics, this method helps structure thinking about a problem. S-curves take almost no additional resources to perform once you have a forecast from another technique. This procedure constitutes the major use of S-curves in industry today.

We are aware of two simple software packages that help in forecasting S-curves, although we do not have firsthand experience with them. It is our understanding that they essentially are curve fitting programs, much like time series estimations. The user supplies known data points and selects possible gamma coefficients to see what curve will result. To really make such programs useful, however, a database is needed of relevant S-curves for many previously matured technologies with which to compare the existing technology. If, and when, such reliable prior data become available, S-curves will be more useful for forecasting.

In the meantime, our expectation for the next decade is that S-curves will not gain in popularity. They will still be used as a theoretical construct to remind forecasters that the trend extrapolation, time series, or regression lines that are forecast must have limits to growth.

Historical Analogies

As George Santayana once noted "Those who cannot remember the past are condemned to repeat it." This warning applies not only to governmental policymakers but also to corporate technology planners. Using historical analogies is a way to follow this advice, while taking a more qualitative approach to trend data than the techniques reviewed so far.

Simply put, historical analogies are used to study historical data from other businesses. Hopefully, such a review will eliminate many mistakes and improve the level of success for your own company. The much touted Harvard Business School case study approach is a form of historical analogy.

In addition to specific case studies, which many companies do examine for relevant information, the largest source of historical analogy information assembled to date is the Profit Impact of Market Strategy (PIMS) database. This database was started at the General Electric Company in the 1970s and is now operated by The Strategic Planning Institute of Cambridge, Massachusetts. PIMS contains over a hundred variables collected from each of over 1700 companies in most industries, over a number of years. Users of PIMS select a company or set of companies that have characteristics similar to their own company (such as having a similar sales volume in the same industry) and examine their performance records.

We know, for example, that most companies today spend between 2 and 16 percent of their sales volume on R&D, depending on the industry. The question is, how much can your company efficiently spend? We examine PIMS for assistance. Suppose a company (or an average of several relevant companies) increased its R&D expenditure level from 3 to 6 percent of sales about five years ago. What happened to its sales, profits, inventory, staffing level, and level of patenting in the five years since this change in strategy? Is there an underlying theory that suggests any change might have been caused by this change in R&D investment? By inference, what does this suggest for your company? These are the types of questions subscribers to the PIMS database seek to answer. This is not forecasting by extrapolating past data into the future, but rather qualitatively forecasting possible futures by analogy to past data.

We have been told by PIMS users that there is much worthwhile information in this database, but for most companies this source plays out after about three years, to the point where the cost of the exploration is not covered by the insights gained. Normal users of PIMS pay approximately $100,000 per year. A potential subscriber should also be aware that The Strategic Planning Institute will expect them to provide data about its own company, for a number of recent years, on all the necessary variables, as a condition of joining PIMS.

The level of training necessary to use analogies is low. Managers already draw inferences from data they collect in news-papers, journals, seminars, and meetings. The use of standardized data sources for this purpose is not a costly transition in terms of staff time. In addition, an analyst or information specialist can be trained to access a database such as PIMS in just a few days to support management's use of analogies. As part of subscribing, PIMS provides the necessary search software to the database.

To use analogies effectively, a relationship uncovered in a case study must make theoretical sense. An inaccurate conclusion may be drawn if an analogous relation is found in several companies, but there is no reason for the result. In other words, cause and effect relationships should be sought in the data.

Analogies can be used in many technology forecasting areas, provided data exist. They might help determine the appropriate level of R&D investment in a technical area, the percent of total R&D to invest in basic and applied research, or the research staffing structure that makes sense for the company. In the areas of forecast-ing technological performance or market penetration, S-curves are actually a specialized form of historical analogy.

Historical analogies only work when relevant data exist. Unfor-tunately, despite PIMS, the amount of previously published publicly available information is not great. This, together with the high cost of access to PIMS, makes the number of corporate users of analogies, compared to the number of users of other trend analyses, relatively small. It is likely that the overall database of analogous corporate information will grow in the next few years together with the number of electronic databases. Such files might even provide a competitor to PIMS in the next decade. If so, the use of analogies will increase somewhat from the relatively minor level of use they enjoy today.

Input-Output Matrices

Wassily Leontif won the Nobel Prize in Economics for inventing input-output analysis (I/O) several decades ago. I/O is a technique in which a regional economy is represented as a matrix of coefficients. A schematic of the main parts of an I/O table are provided in Figure 7.

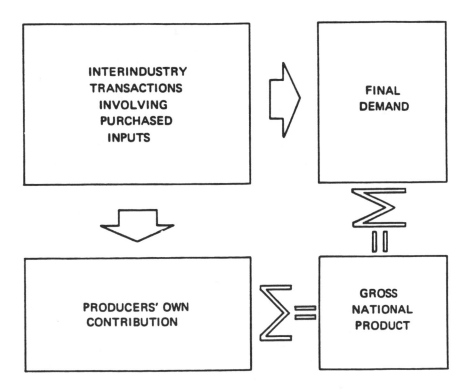

FIGURE 7.   MAIN PARTS OF AN INPUT-OUTPUT MATRIX

In the upper left corner is a matrix of interindustry transactions in which the columns of the matrix indicate the amount of goods and services a sector of the economy must purchase from other sectors of the economy to operate. The rows of the matrix indicate the level of sales by one sector of the economy to all other sectors of the economy. For example, one cell of the overall matrix will indicate the dollar value of goods and services that the electrical machinery industry must buy from the trucking industry to operate.

There are one or more columns of final demand, which measures the amount of production that is consumed, rather than sold to other sectors of the economy. One row across the bottom measures the amount of its own production that each economic sector consumes. Finally, note that the sum of the bottom row, or the sum of the final demand columns, is the usual measure for gross national product. Once the final demands have been specified for a particular economic region, the model is solved using matrix algebra, such that all the interindustry transactions are balanced.

The degree of disaggregation of the economy into sectors varies from one I/O table to another, but it is usually between 100 and 400 sectors. I/O tables are usually constructed for national economies or subnational regional economies. New York University, in concert with Brandeis University, has constructed a global I/O table by interrelating I/O tables for most of the industrial countries in the world. (This is a more modern global model than the one created by Forrester and Meadows in 1970 using systems dynamics.)

In technology forecasting, the primary use of an I/O table is to measure the effect that a change in technology will have on the economy. As an example, it might be useful to know the overall economic effect of a vast increase in the conversion of shale oil to marketable crude oil, or the economic changes that will occur when a large number of inexpensive superconductors are available. I/O is useful for the macroscopic questions that arise in the evaluation of a business technology strategy.

I/O is not useful for middle- to long-term forecasting because the table represents the status of the economy in a particular recent year. Nor does I/O allow forecasting technological performance characteristics. It is best used only for measuring the effect of a change in technology on the entire economy.

In the U.S., publicly available tables may be purchased from several government agencies, including the National Bureau of Economic Analysis in the U.S. Department of Commerce. Tables for other countries are usually available from a similar government agency. One drawback to the technique is that these tables of coefficients are often several years old, and the technological relationships may be obsolete by the time they are available for analysis.

Software to analyze a table usually must be custom developed, but can probably be written one time for under $30,000. Consulting firms, such as Battelle, already have such in-house software and I/O tables. Battelle usually charges $10,000 to $20,000 to conduct a complete I/O analysis for a client. This is an indication of the level of effort necessary to utilize this method in-house.

Many state and federal government agencies conduct I/O studies, as do companies such as Exxon, Monsanto, and Westinghouse. (Publicly available case studies may be purchased from the U.S. National Bureau of Standards.) However, because they usually provide a focused answer to a discrete type of problem, I/O studies are not widely used by companies. Our expectation is for a continuation of this trend. I/O will be used in the 1990s by relatively few companies to solve the particular type of technology forecasting problem outlined in this section.

The eight techniques reviewed so far are applicable to needs beyond technology forecasting. They are also used for many other types of forecasting. We next consider three methods that are primarily used for technology monitoring, scanning, and planning. All are special forms of bibliometric analysis, which has its roots in the 1940s.

One good way to acquire business intelligence about the technological developments occurring in a competitor's research lab is to directly ask the researchers about their work, but you may collect inadequate or inaccurate information. You may also not know who to ask. Another way is to examine the researchers' published output. This can involve examining their patents, their journal or book publications, or their conference presentations.

Patent Trend Analysis

A particularly useful tool for technology monitoring is patent trend analysis. Companies file patents on technological innovations to protect them legally from copying by competitors. They often file patents concurrently in the United States, West European countries, and Japan to protect their intellectual property worldwide. These patents are a matter of public record. Therefore, an analysis of patents provides information on technology trends and actors in the development of new or improved products and processes.

Battelle follows a six-step patent trend analysis method. A similar method is followed by most companies. Because this technique is new and not well published, the details of the six-step method are provided below:

1 **Define Study Objectives.** A topic must be defined based on specific technology management concerns. Technologies of interest can then be focused upon, based on the questions the analyst seeks to address and the ultimate use of the information.

2 **Establish Problem Domain.** Key technical issues or problems should be understood to build a framework for the analysis. A patent categorization scheme should be developed to permit more detailed and insightful analysis of the patent data. It is frequently useful to begin a patent analysis with a review of nonpatent sources of information on the topic of interest. These sources can include technical journals, business magazines, internal company reports, etc. Such a preliminary review can serve several important functions. First, this review can be helpful in the search for the relevant patents and their subsequent categorization. The data may also reveal corporate affiliations or subsidiaries that may not be generally known. Finally, nonpatent data can supplement and deepen the insight into the problem domain as revealed through the patent indicators, by connecting changes in the behavior of the indicators with information from other data sources. Patent trend analysis should not stand alone. It should be viewed as only one input into the decision-making process. Other sources of information both enrich and clarify the interpretation of the patent data.

3 **Obtain Relevant Patents.** A search strategy is developed for identifying patents using commercial patent databases.

It is important to have data from as many years as possible. There are two principal reasons for this. First, while frequently only recent trends are of interest, having a limited number of years' data does not allow calibration of the indicators in a specific area. Second, a limited number of years' data can affect the behavior of the indicators, particularly those derived from patent reference data.

Because a patent is an exclusive 17-year lease on a particular technical solution, it is prudent to cover at least the last 17 years, unless the technology has emerged more recently.

Searching for the relevant patents usually requires the use of a commercial patent database. The search process is more an art than a science. Several basic strategies can be used, either individually or in combination:

- Key word
- Patent Office classification
- Citation data
- Abstract review
- Full text review.

Each of these techniques has advantages and disadvantages. An experienced information specialist knows how to combine searching techniques to mitigate the disadvantages. Even though patent searching is often tedious and time-consuming, it cannot be stressed too strongly that the quality of the output is dependent on the relevance and completeness of the group of patents selected for analysis.

4   **Load Patent Data into Software.** Data obtained from searching commercial databases are loaded into a computer file. The data are edited and each patent is categorized to aid analysis.

Additional insights can be gained by categorizing patents according to some relevant dimension such as type of technology, type of material, method of manufacture, process versus product, or organizational affiliation (e.g., central research versus product/division laboratory). Then patent indicators can be calculated for a subset of the patents. For example, after patents are classified into technical categories, activity graphs can be used to show changes in emphasis over time. One area of the technology may receive most of the patents for a period of time until a breakthrough or other circumstances motivate a change in focus (see Campbell and Levine, 1984).

5   **Produce Computer Output.** In this step, specific measures of patent trend behavior are generated. Subsets of the

data are analyzed to gain greater insight into technology trends and competitor capabilities.

Generally a software tool is used during Steps 4 and 5 to make the patent trend analysis process manageable. Tasks usually performed by software include the following:

A   Downloading and merging patent data from commercial patent databases into a personal computer

B   Editing data into files to be analyzed

C   Eliminating database errors, such as duplicate patents and misspelled inventor or company names

D   Calculating analysis indicators

E   Providing tabular and graphical outputs, as well as electronic spreadsheet output.

6   **Interpret Analysis Results.** The key patent indicator results are finally combined with other sources of business and technical information to prepare an implications assessment and final report.

The analysis indicators facilitate organizing and interpreting large amounts of patent data by extracting meaningful patterns. These indicators have been applied by many companies in numerous studies to help identify areas of technological growth and relate the information to needs for planning and development. Some important analysis indicators that can be computed from patent filings include the following:

Activity analysis provides a frequency count of patents by technology type year by year. It indicates the level of inventive effort over time so the analyst can identify increases and decreases in emphasis in specific technical areas.

Dominance is a measure of intercitation between the key companies in the sample. This reveals who has cited whom for related patents. The most frequently cited companies are usually the ones that have established strong patent positions in a given technology. For example, Figure 8 gives the dominance matrix for sodium-sulfur batteries.

Company characteristics provide further information about the patenting behavior of each company. They can be used to rank companies according to various patent characteristics, such as number of patents, number of inventors, and average patent age.

Portfolio analysis supplies tables of patents held by each company, separated by granted patents and published patent applications.

Specific company activity allows detailed analysis of a company's patent activity, including indications of changes in R&D staff.

To corroborate patent findings based on these indicators, sources of additional data should be reviewed including SEC filings, annual reports, product catalogs, trade publications, or even phoning the important inventors and asking questions about their work. Several such sources of data are now available in electronic files on databases offered by vendors including Orbit and Dialog.

Patent trend analysis often uncovers very interesting new technologies that might present seemingly unrelated business opportunities. For example, can your company benefit from laser driven sensor systems that actively alter automobile suspension performance, or high resolution large-screen television systems, or biodegradable plastics that eliminate trash? Are there actions you can take to profit from the introduction of new technologies?

Patent trend analysis has been used increasingly by companies all over the world during the last five years. Users include 3M, Kimberly-Clark, British Petroleum, Weyerhaueser, Du Pont, Hitachi, Sumitomo, and IBM along with dozens of others. There are hundreds of companies performing patent monitoring and limited analysis of patent holdings. Large companies in all countries that work in any of the patented fields (electrical, chemical, mechanical, and recently biological) are using patent databases to varying degrees. And this use is accelerating rapidly as patents become a major source of international competitive intelligence about technology. In addition, in recent years more companies are seeking to protect their intellectual property, so the number of patent filings of important inventions contained in the databases is increasing.

Many of the basic patent trend analysis techniques were developed by Battelle between 1979 and 1983 under a grant from the U.S. National Science Foundation, and Battelle can provide nonproprietary case studies using patent trend analysis. After a decade of development it seems that the Japanese are leading in their analysis of

CITATIONS RECEIVED \ REFERENCES GIVEN	Ford Motor Co.	U. Utah/Ceramatec	U.S. DOE	Dow Chemical	Dupont	Corning Glass	General Electric	EPRI	Toyota	Ind. Science - Tech.	Yuasa Battery	Chloride Silent Power	Electric Council	Sel/State Industry	British Railways Board	Brown Boveri DeCie	Generale d'Electricite	Other Inset	Total
Ford Motor Co.	57	4	1	19	2		22	6	9	7	4	12	15	10	6	14	18	17	223
U. Utah/Ceramatec		2					2										2		6
U.S. DOE																	1		1
Dow Chemical	3			26	8	14	2	4	1	2	1	4		2	3	7	1	7	85
Dupont						1												3	4
Corning Glass																			
General Electric	18	2					24	16	1		1	8	1	1	3	3	10	9	97
EPRI	8	1					1	14				4			5	1	2		36
Toyota	1	4	1				2	2	1			5	5	1	1	1		8	32
Ind. Science - Tech.	2						1				1	3	2	2	3	1		1	16
Yuasa Battery	3						2	1	1		1	5	5	4	2	3	1	3	31
Chloride Silent Power	12	1	3			1	5	14				20				9		10	75
Electric Council	8	4	2	2			1	3					17	4	4	1	2	2	51
Sel/State Industry	5						7						7	1	2	2	1	4	35
British Railways Board	12						2	2					4	1	1			2	24
Brown Boveri DeCie	2												1			6		6	15
Generale d'Electricite	7		1				11	2		2	1	6	3		1	4	1	5	44
Other Inset	4	6	4	1	2	6	4	4	1				10		3	5		31	78
Out-of-Set	61	6	8	38	11	3	109	36	15	6	2	48	6	6	9	28	20	81	493
Total	203	21	27	88	23	24	186	111	30	19	10	181	43	35	27	96	61	194	1348

Source: Adapted from Ashton and Sen, 1989, p. 40. Reprinted with permission.

FIGURE 8.   PATENT DOMINANCE MATRIX FOR SODIUM-SULFUR
BATTERY TECHNOLOGY

patents, European companies are in second place, and North American companies are trailing in their use of this method.

Over the last ten years, patent databases have become more extensive and easier to search. Personal computers provide new access to the databases and make it easier to use the data. Two software products for analyzing downloaded patent data commercially available today—PATSTAT Plus by Derwent and PATENTS-PC™

by Battelle. Both make the job of analyzing hundreds or thousands of patents more manageable. In addition, many companies have created their own proprietary software for analyzing patents. Some are doing their analysis using commercial database products, such as DBASE III, in a limited way.

Most of the dozens of large companies working in this area have staffs of one to ten people devoted exclusively to performing patent analysis, with a mode of about two people per company. These researchers have either been trained or have trained themselves over a period of months or years to know necessary details about the patenting process in most countries, the available patent databases, how to search for relevant patents, and how to conduct a meaningful patent trend analysis.

The major cost in conducting a patent trend analysis is these staff resources. Next is the $1,000 to $5,000 cost of searching for patents for each technology studied. Computer and software costs are relatively minimal over the long term, amounting to about $10,000. Overall, the cost of patent trend analysis is one of its major disadvantages compared with the other trend analyses reviewed above.

Other drawbacks to patent trend analysis are that not all inventions are patented, not all patents are useful or commercially meaningful, and there is at least an 18-month delay (sometimes up to 36 months) between the time the research is performed and the patent filing appears in the database. For most industries, the indicators do not provide insights into what will happen more than three years in the future. Thus, it is necessary to update or repeat the patent trend analysis every 6 to 12 months for maximum advantage.

Despite these disadvantages, the technique still holds great promise as a way to learn about competitors' R&D work, potential new products, and manufacturing processes. Battelle has conducted studies that show that patent databases can yield predictions of future marketplace developments 6 to 18 months before they actually occur. It is now one of the better ways to monitor technological developments around the world. It is also advantageous for uncovering candidate firms for acquisition or for technology licensing. Some use the method to identify key inventors that they should consider hiring.

Based on the trend to date, it appears patent trend analysis will be even more widely used during the next decade. It is expected that patent databases with continue to improve in scope, coverage, and ease of access; that better generalized analytic software will be published; that more powerful personal computer hardware upon which better analytics can be performed will soon be available; and that more uniform patenting rules around the world will come into being. The introduction in 1983 of the European Patenting Office has already resulted in standardizing the data from European countries, and even more standardization is expected from the 1992 cooperation agreement. Both the Japanese and the U.S. patenting offices are contemplating changes in procedures that will standardize world-wide data further. All these trends should make the use of patent trend analysis easier and the interpretation more consistent.

Scientific Literature Analysis

A complementary method to patent trend analysis is scientific literature analysis. The process is much the same, with the central difference being that the data analyzed are not patent filings of inventive activity, but rather data on the publications of researchers explaining their inventive activity.

The procedure for scientific literature analysis involves the same six steps as patent trend analysis, namely:

1   **Define the technology area** to investigate.

2   **Establish the problem domain,** providing details on which year or years of publications are to be studied.

3   **Search** all scientific and technical publications for relevant articles, possibly restricting the search to articles in chosen languages.

4   **Load relevant data** about these articles into a database, including author names, article titles, possibly an abstract of each article, references given in the bibliography, etc.

5   **Analyze the database** to determine the number of articles published by author, by country, by technical area; to determine which articles have subsequently been cited in the most bibliographies, as an indicator of the importance of each article; and to determine which companies are publishing the most articles in the technical area, as an indicator of research effort. Normally Steps 3, 4, and 5 are

performed using computers, to reduce the human effort to a manageable level.

6   **Analyze the implications** of these indicators and make business decisions.

The advantage of this technique is additional insight into the technologies being developed in research labs, what companies are doing this work, the key researchers (authors), and which new firms are likely to emerge as champions in a given technological area. In many cases, articles are written and published much sooner than patents covering the same invention are issued—thus the scientific literature database is somewhat more current. However, studies indicate that only one-third of the patented research activities have ever been published, so the scientific literature is not as comprehensive as patents. It is not yet clear how far into the future insight can be gained from scientific literature analyses, but it is guessed to be about three to four years. Beyond this, the insights will be obsolete. Thus the technique must be used on a fairly regular basis, at least annually, to provide maximum advantage.

The main consulting firm performing scientific literature analysis today is the Center for Research Planning (CRP) in Philadelphia. Some of the original development work was performed under a U.S. National Science Foundation grant. CRP has also consulted for a number of industrial and government clients. Generally, CRP charges between $50,000 and $150,000 per study. A few companies perform their own variations of this method as an additional form of technology monitoring. There is one electronic database containing the data on scientific literature necessary to conduct this work. It is relatively expensive to access. CRP leases access to its analysis software; however, there are no commercially available software alternatives.

This method has good potential for technology monitoring. It is expensive today; but if competition increases, the method can be expected to decline in cost. The question is when will this occur? Based on an S-curve analysis comparing the introduction of scientific literature analysis with the introduction rate of the previously developed patent trend analysis, we can expect competition to develop in approximately 1992, once the method has caught on in a number of companies. Thus our forecast is for increased utilization of this method during the latter half of the 1990s, after costs have decreased significantly.

New Trend Analyses Techniques

The overall forecast for the future of the ten trend analyses tools mentioned above is that they will continue to be used, although less by themselves and more in conjunction with expert opinion and multi-option analysis methods. Some, such as S-curves and to a degree econometric models, will decline in use. Others, such as patent trend analysis and scientific literature analysis, most likely will greatly increase in use.

In addition to these existing techniques, a new trend analysis method will be developed during the 1990s. This technique will come about because of the combination of three already developing trends in computer hardware, computer software, and data:

1   Hardware for mass storage of up to 1000 megabytes or more of information is coming soon in the form of Write-Once/Read Many (WORM) optical disk systems, Compact Disk Read-Only Memory (CD-ROM) database disks, etc. Already available is the removable WORM cartridge that stores 786 megabytes offered by Storage Dimensions. With the technology, over 250,000 pages of text covering a particular topic can be stored on a single cartridge; users can simply use one removable cartridge per technology that is being monitored or forecasted. In addition, scanners will improve in resolution and speed during 1989.

2   Software that works with scanners is rapidly improving (such as Calera Recognition Systems' Truescan), meaning that information on printed pages with widely variable fonts, spacing, pictures, etc. is readable as an actual text file that can be used in word processors. Also, software that is now being introduced searches these scanned files for keywords, phrases, etc. This allows users to more easily perform an electronic hunt for key company names, for key product types, and for key technologies in the making, and note these for a human to review. Hypertext, which allows searches by strings of words and the interlinking of documents or segments of documents with other documents and electronic data files, is expected to gain rapid market acceptance. Software using the hypertext concept is now being written. Finally, programs (such as Persoft's Ize) that allow for easy organization of vast amounts of seemingly unrelated information (by letting the user place

her/his own key topic indices on each document that was first scanned into a database and second deemed to be relevant information after a search) will allow access and recall of the scanned results for synthesis and technology monitoring.

3   Electronic databases are growing in number and complexity. The content is still highly variable, but the different databases are becoming more consistent and searchable. The amount of data available in these files is also increasing as time goes by, obviously, because the old databases are being continually updated. As fiber-optic networks expand globally, access time and cost for using electronic databases will continue to fall. Finally, with the ability to scan in your own data sources, there is now almost unlimited data available to search. In fact, we expect that the number of commercial databases will expand as these scanning tools and search tools are used by new/existing vendors to provide new organized data for subscribers. For example, it soon will be economically possible for a database firm to scan in newspapers from all over the globe, sort the articles by topic, add an index to the result, and provide the entire database on CD-ROMs on a monthly basis. This will radically change the way technology scanning is performed by many companies.

In combination, these evolving technologies will allow each researcher to construct her or his own database of very current material covering any technology. The source of data will be anything that is in writing, found anywhere in the world. Software will allow this information to be categorized in a meaningful way (exactly as in patent trend analysis), synthesized, and used for business decisions.

In addition, the availability of specialty software that allows analysis of commercial electronic databases has also increased in recent years. For example, scientific literature citation analysis software is now leasable, and two different patent trend analysis software packages have been published in the last two years. Software products such as these will continue to be introduced. Generalized software for analyzing user-created WORM databases cannot be far behind.

It is worth noting that the number of specialists working with such databases and search software is growing rapidly. For example, at the 12th International Online Information Meeting held in London, December 6–8, 1988, there were 6,372 attendees, an increase of 27 percent over the previous year's attendance. Another insightful statistic comes from the U.S. where the membership of the Society of Competitive Intelligence Professionals (SCIP) has doubled in size every year since its formation, now numbering approximately 850. Many of these professionals are monitoring, scanning, and forecasting technology changes within their own companies and changes involving outside competitors.

Conclusions

For a moment, we should review the two assumptions underlying all trend analyses. Recall the basic assumption that the future is a continuation of the recent past. There is also the assumption that human behavior follows natural laws; thus, such laws can be expressed quantitatively. The other assumption is that there is one future and it is predictable if you understand these underlying laws.

In fact, each of these assumptions is doubtful. Seldom does the future follow directly from the past. If it did, a simple trend extrapolation would be fairly accurate as a forecast. But it is precisely because such simple extrapolations usually are not accurate that regressions, econometric models, systems dynamics, and S-curves were invented. These latter techniques recognize that the world is more complicated than a simple forecast allows.

In human nature, there are no constants, just some variables that change more slowly than other variables. These slowly changing variables represent the constants in a trend analysis or econometric model. This situation leads to more inaccuracy in forecasting social conditions (which determine product sales) than in forecasting physical phenomenon, where some natural laws do seem to hold constant (such as Planck's constant, Kepler's laws of gravity, or the speed of light).

In light of this, one complaint often raised is that forecasts made with trend analyses are not very accurate (for just one example, see Brody, 1988). In fact, all forecasts involving human behavior are subject to far more error than forecasts made in the physical sciences. However, several recent studies of the accuracy of trend analyses forecasting suggest that forecasts have been improving as techniques and data have improved during the 1980s (for examples, see Faruqui, 1987, or McNees, 1988). It also appears that models that forecast macroscopic issues tend to be more accurate than models that forecast microeconomic problems—possibly due to the larger structural changes necessary to jolt a macroscopic forecast.

The assumption of a single predictable future holds two dangers. First, as we have seen, while an effort is made to predict the future, it is unusual for any particular forecast to be accurate. This has led to such practices as averaging the predictions of several forecasts to create the "best" forecast. Second, there is probably more than one possible future. Actions taken today will affect the future. Heisenberg suggested that measuring information about a physical process would of itself affect the results of the measuring experiment. So too does technology forecasting affect business decisions that in turn affect the forecast. Managers must recognize that a range of possible futures do exist, and then monitor the path actually being taken as the forecasted time point is approached.

Despite these problems, trend analyses are likely to remain a major technology forecasting tool for companies through the 1990s. They offer a relatively inexpensive way to obtain a better educated guess about the future.

There is much merit in examining where we are today, for this gives major clues about tomorrow. The past provides insights into the future inasmuch as it establishes what is normal or usual or even what is possible. While the past does not determine the future, because major changes in human relationships do occur, it does set boundaries on future outcomes that require a major new circumstance to change. This suggests that trends will more likely hold in the short-term than the long-term.

Our view is that trend analyses are a very important set of tools, but they may be overemphasized and overrated—particularly when one considers that the assumptions upon which they are based seem to empirically make them about as accurate or inaccurate as expert judgment. Too much emphasis is placed on accuracy of the methods, and not enough placed on learning about the area being forecasted.

The greatest use of trend extrapolation, time series estimation, and regression analysis will continue to be forecasting the short term, say the next two years. Patent trend analysis and scientific literature analysis provide some insight into technology changes that will occur over one to four years. Some of the more complex methods, such as systems dynamics or S-curves, are used for long-term technology forecasting, from two to fifty years into the future.

There are reasons other than forecasting to use these tools. Patent trend analysis and historical analogies are the best of the trend analyses methods for identifying potential technology strategies that a company might follow. Among the trend analyses methods, historical analogies and econometrics are best for evaluating the selected strategies—thus determining the effect of actions taken today on the forecast.

In sum, the major drawback to trend analyses is the underlying assumption that the future is a continuation of the past. The social aspects of human nature are far too complex to allow this to occur. Trend analyses typically are very thorough and rigorous in microscopic detail, but vague in macroscopic scope. Thus a major need for the 1990s is to combine thorough microscopic trend analyses with expert judgment and multi-option techniques, to provide the best possible forecast, together with a greater understanding of the technological area itself.

C H A P T E R

3

EXPERT JUDGMENT

Since man's earliest conscious wonderings about the future, no method of thought has been relied upon more than expert judgment. For thousands of years, the "experts" were extraordinary humans who claimed divine or mystical insights: prophets, oracles, seers, sages, sibyls, and fortunetellers. Their expertise, most often, was in their ability to be convincing rather than accurate. In modern times, humans with all of their scientific sophistication have renounced the occult and have turned to those with knowledge and reason: scholars, specialists, consultants, and "experts." Although the source of foresight may differ, judgment today as in ancient times is still an ill-defined combination of information and intuition.

Expert judgment, to define it the best we can, is the assertion of a conclusion based on evidence or an expectation for the future, derived from information and logic by an individual who has extraordinary familiarity with the subject at hand.

To one degree or another, all methods of forecasting and analysis involve expert judgment, whether it is one person's or a group's, whether it is expressed in numbers or in words. As Rand analyst E. S. Quade observed about 20 years ago, "Intuition and judgment permeate all analysis. . . . As questions get broader, intuition and judgment must supplement quantitative analysis to an increasing extent." Expert judgment, therefore, becomes particularly important in the analysis of highly uncertain and complex topics, such as the future.

Of all the forecasting methods, the one that high-level corporate managers understand the best and practice the most is expert judgment, particularly their own. Many successful business people trust their intuition, which in many cases had to have worked or they would not have been successful. On the other hand, as much as they may trust their own expert judgment, corporate managers have become very skeptical of other people's expert judgment. They rightfully demand to know why experts forecast as they do. Expert forecasters of technologies in the corporate environment must be convincing based upon information and logic rather than on divine inspiration.

Professor Joseph Martino in his widely read textbook, *Technology Forecasting for Decision-Making,* asserts that expert judgment is called for when

1   Historical data, which is so vital for trend analysis, does not exist, so that judgment becomes a surrogate for trend extrapolation,

2   The effects of external, changing factors appear to invalidate the results of trend extrapolation based on historical data alone, and

3   Ethical and moral (and perhaps political, too) factors are sufficiently important to override strictly technical and economic factors.

We agree with Professor Martino (indeed, with his kind indulgence we have somewhat reworded his criteria into our own language), and to his list we would add further situations for expert judgment:

4   Data may be available, but may be very difficult and expensive to process given the total circumstances and the utility of the technological forecast;

5   Data are not available from printed or electronic sources, so they can be gained only from experienced experts and specialists;

6   The interrelationships of many factors and their complicated and cause-and-effect interactions are very important and may change the projections of any one factor;

7   The behavior of the "experts" themselves may impact the result of the forecast, such as when the "experts" are consumers asked to express their reactions to whether or not they might buy a technological product.

Expert judgment methods are certainly not unique to technology forecasting and strategy analysis. They are general, domain-free methods applicable to virtually any topic. Applied to technology forecasting, they are most likely to be used early in the R&D process, when trend and statistical data are the rarest and the least reliable. Furthermore, expert judgment is used in conjunction with virtually all other forecasting and strategy analysis methods.

In the discussion below, we will cover the following methods of expert judgment used by companies in technological forecasting for new product and process development:

- Interviews

- Questionnaires

- Group Dynamics
    —Delphi Method
    —Idea Generation
    —Nominal Group Technique.

Interviews

Interviewing is a well-known and often practiced technique of gathering information. We have all witnessed TV interviews ranging in skills from the local news to *Sixty Minutes*. Most researchers have at one time or another interviewed other researchers and have been interviewed. Some interviews are formal and structured; others are very casual and free flowing, much like the exchange of information at a cocktail party (where, by the way, the Soviets typically gather much intelligence unbeknown to the courted interviewee).

The purpose of interviewing is to gain the in-depth judgment of an expert about the forecasting topic. The interview goes beyond the more limited and structured form of written expert judgment that is available through a thorough literature search. If you knew or trusted just one individual to give you your forecast, then just one interview would be necessary (as was the case with oracles and fortunetellers who allegedly knew "the one and only" future). Otherwise, to supplement the always finite knowledge and human biases (or "convictions") of individual experts, you will need to conduct and synthesize numerous interviews of experts.

A few basic practices should be followed for successful interviewing, such as the following:

1 The interviewer needs to give thought to whom he or she wants to interview and why. Interviews of experts should not be planned and executed carelessly. The types of information needed should be identified first, and then the names of people most likely to supply that information should be compiled. The number of and time available for the interviews depend upon the amount of time and money available to the interviewer. Another factor for budgeting resources is the importance of the information for the purpose at hand. Furthermore, questions should be written down in advance to capture the information needs of the interviewer.

2 The interview can be conducted in person or by telephone. Typically, an interview means the personal exchange of questions and answers; if the questions were written, the instrument would be a questionnaire rather than an interview. Longer interviews should be conducted in person, shorter ones over the telephone. Face-to-face interviews have several advantages: the interviewee is freer to respond in his or her own way, unplanned but potentially significant digressions can be explored, additional information in the form of facial expressions and body language can be gathered, and a personal rapport can be cultivated. Research and interviews are potentially forms of marketing and should be treated with politeness and seriousness. Disadvantages include the time and expense of interviews, especially if travel is required (and the experts may indeed be spread out from Tokyo to Torino). In the circumstances of having to do many short interviews over considerable distances but with not much time, the interviews obviously should be conducted over the telephone.

3 The time and place of the interview, whether in person or over the telephone, should be coordinated with the interviewee to facilitate his or her full cooperation. A letter explaining the purpose of the interview, and perhaps a sample of questions, should be sent in advance to the interviewee.

If the interviewee does not respond within ten days or so, then the letter should be followed by a telephone inquiry.

4   Always arrive when planned and telephone when arranged. Exhibiting bad manners is a bad research technique and poor marketing.

5   Ask your questions in your way and let the interviewee answer in his or her own way. Listen to what the interviewee says, not what you are expecting to hear. Like a rigorous scientific experiment, the interviewing process should be a fair and realistic gathering of information without the interviewer disturbing the results (hence, the Heisenberg uncertainty principle applies to social scientific research).

6   The interview method is probably best applied when you have identified an individual with rare information and insight, when a corporate political need exists to include certain individuals in the research process, or when you are carefully cultivating the interviewee for business beyond the immediate scope of the interview.

Virtually all corporations and analysts doing technology forecasting and analysis have used interviews to gather information. Typically, interviews are used as an input to one or more forecasting methods rather than as a forecasting method per se. In unusual circumstances, interviewing may be used as a forecasting method, such as when the interviewee is an important player whose actions may determine the outcome of the forecast topic. At Battelle, we have used interviews to gather expert judgment and specialized information for scenario analysis, trend analyses (especially trend extrapolation and patent trend analysis), and relevance tree analysis. We are not aware that interviewing is applied in connection with many forms of statistical forecasting methods, S-curve analysis, and portfolio analysis (although it certainly could be).

Japanese companies are particularly adept at the interview method. They seek out many sources of technology and business information, and they seem to prefer face-to-face discussions over impersonal surveys. American companies, in contrast, are likely to use questionnaires and statistical analyses more than the Japanese. European companies appear to do more interviewing than the Americans, but less than the Japanese.

Questionnaires

Questionnaires are generally interviews prepared as written questions to which the respondents reply without the presence of the interviewer. It is an impersonal and multiperson way to survey expert opinion. One advantage of the questionnaire is that you can survey many more experts through a questionnaire than through interviews. A significant disadvantage of the questionnaire is that the structuring of questions and answers keeps respondents from saying exactly what they think.

Much has been written about questionnaires that precludes the need for another detailed discussion of them here. (In particular, see Seymour Sudman and Norman M. Bradburn, 1982.) Based on our experience with questionnaires for technology forecasting, we suggest the following guidelines:

1   Use questionnaires as a means to gather expert opinion and data, but not as a forecasting tool per se (in much the same way as interviewing). We would use them as inputs to other forecasting methods, such as scenarios or trend analyses.

2   Select participants carefully to assure participation. Ideally, you should know all the participants and their particular areas of specialization. Because this is usually not the case, you should use proven mailing lists of the kind of experts you need, typically members of associations, organizations, and groups with special interests and expertise. For example, members of the Planning Forum would be well qualified to respond to questionnaires about academic and corporate planning practices. You should know your participants by group identification if not as individuals. This knowledge about the correctly selected target of the questionnaire leads to meaningful results more often than a careless "shotgun" approach.

3   Determine the kind of information you need and why you need it before you structure the questions. Let the purpose direct the structure.

4   Keep the questionnaire as short as possible. The shorter it is, the more likely the respondents will complete it fully and return it. Many questionnaires are too long: they ask too many questions requiring too much data. After you

have determined the data you need, keep the questions focused on that goal and do not ask extraneous questions.

5   Structure the questionnaire, but give the respondents the opportunity to express their own views. Not all questions should have answers of "yes" or "no;" indeed, not all questions should have three, four, or five choices of answers. Some questions should be "essay-type" in the sense that the respondents use their own words to express their own thoughts. At the end of the questionnaire, you should allow space for the respondents to put down their own questions and answers to supplement yours.

6   Make the appearance and mechanics of the questionnaire as "user friendly" as possible, as though you yourself had to complete it.

The principal advantage of a questionnaire is that you structure the questions according to your needs for information. This "management" of the resulting data makes it more compatible for your style of analysis and more "usable." The second benefit of a questionnaire is that you can survey a large number of people, from a few dozen to hundreds (even thousands if you had the time and money). More data often make some analysts feel more confident in the "validity" of their analytical results. A third benefit is that the questionnaires can be reviewed repeatedly and saved, so that no content is lost (which so often happens with interviews that are not taped).

The principal disadvantage of questionnaires is that they are frequently costly and time consuming. Questionnaires require time and money to compose and refine, to send out, and to tabulate the answers. In addition, certainly not all of the recipients will in fact respond; a 75 percent response rate is excellent and often even 25 percent is acceptable. Therefore, time and money are wasted on questionnaires that go unreturned. It is necessary to decide whether the value of the results is worth the effort of attaining them. However, this is not only true of questionnaires, but of all data gathering and forecasting methods.

Another disadvantage of questionnaires is their structure. Yes, you get to ask the questions that you want answered, but are you asking the "right" questions? Questionnaires can be misleading, confusing, and even irrelevant. To address this problem, space is needed to allow the respondee to add comments (even though free-flow remarks are difficult to tabulate and integrate into the questionnaire analysis). The questionnaire should be tested on a sample group before using it with everyone.

Questionnaires have become very popular with marketing-oriented companies, especially consumer product and entertainment companies. Companies that like to think of themselves as "market-driven" tend to do surveys of customers, either directly or indirectly through specialized survey services, through mail questionnaires or quick interviews. In most of these situations, the survey is truly an opinion poll: Would you buy product "x"? How much would you be willing to pay? Do you like the color? Etc. The person surveyed is being asked to predict his or her own behavior—the person is used as an expert on himself, herself, or a represented group.

As a tool of market forecasting, rather than as a tool of technology forecasting in its narrow definition, questionnaires are used extensively in the middle and late stages of R&D. They should be included in any combination of tools used for "product forecasting."

In terms of accuracy, questionnaires are predictive when they are addressed to people who can affect the result of the forecast topic. These are actually intent surveys rather than prediction surveys—they are not asking people whether a new product, for example, will be successful or not (let alone what sales levels will be), but whether people would or would not be interested in buying it.

Questionnaires are used extensively in technology forecasting for establishing probabilities for technological performance and dates by which specified performances will have been achieved. These probabilities should be seen for what they are—truly hypothetical. Yet they reveal the expectations of the experts whose opinions they reflect. Again, questionnaires are best when they measure the intentions, expectations, and even attitudes of those surveyed. One of the great shortcomings of questionnaires in general is that they rarely ask the respondents why they answer as they do. Such information, although difficult to compile and analyze statistically, could be very valuable for corporate planning to address "expert" (and consumer) intentions, expectations, and attitudes. In this respect, planning and marketing surveys (both oral and written) have fallen behind the craftsmanship of political polling.

We know of no example where a corporation is relying upon questionnaires alone to forecast technology or to plan R&D. We know of many examples where corporations have used a questionnaire at some stage of forecasting and planning, typically as informational input and market assessment. As surveying tools become more sophisticated, we expect corporations will use questionnaires more in the 1990s than in the past, but for much the same reasons as in the 1980s.

Group Dynamics

As an alternative to interviews and questionnaires, you can convene groups of experts to express their views collectively. Often a synergy of creativity occurs with experts working together that does not occur when they express their views alone (as they do in interviews, where they interact with the interviewer but not other interviewees, and with questionnaires, where they typically do not know how other experts are responding to the same questions). Groups, however, can be difficult to manage, so a disciplined process is required to achieve success.

We define "group dynamics methods" as procedures used to solicit judgments from an assembled group of experts in a systematic way toward achieving a goal. We will specifically discuss three group dynamics methods: the Delphi Method, Idea Generation, and the Nominal Group Technique.

The Delphi Method. The Delphi forecasting method was developed by the Rand Corporation in the 1950s for the U.S. Air Force. The concept was that an iterative questionnaire of experts would produce a consensus and accurate forecast when direct information for trend analysis and prediction was not available. Hundreds, if not thousands, of Delphi projects over the last 30 years have demonstrated the validity of the first part of the concept, but not necessarily the second. Delphi will produce a consensus forecast, but it has not been significantly more accurate than other forecasting methods. Indeed, for all of its rigor (and expense), it suffers from the same disadvantages as other forms of expert judgment forecasting: biases of over-optimism, biases of over-pessimism, incomplete information, lack of synthesis of many trends to reach a comprehensive view of the future, and lack of imagination to consider structural changes that can radically change the flow of trends.

The fundamental steps of the Delphi method are simple, although many variations have been attempted. The basic steps are as follows:

1   The Delphi managers determine study goals, requirements, and then **structure the questionnaire** accordingly.

2　The Delphi managers **put together a list of experts** to complete the questionnaire. The experts do not have to be convened as a group, but they can be if circumstances permit. The number of respondents can vary from dozens to thousands, although a few hundred is a practical limit. The managers typically mail the questionnaire with instructions and then wait for the responses. They do not at this stage inform the respondents who the other experts are.

3　The managers **receive and tabulate the responses** to the questionnaire. The responses are tabulated and the results are sent back to the respondents along with a list of names. To qualify as Delphi, the process should be iterative. The answers to the first round and the names of the experts certainly do influence the judgments made on the second round. Typically convergence occurs; and the more iterations made in the administration of the questionnaire, the greater the degree of convergence of opinion. The number of iterations varies greatly, from at least two to as many as six, although three or four seem usual. The iteration and the feedback information with the goal of consensus make Delphi a group dynamics method rather than a questionnaire technique as discussed above.

4　The result of the Delphi method is a **consensus forecast.** Delphi has been used extensively for technology forecasting, although not for market surveys. It has been used for

- Identifying applications for existing or emerging technologies

- Establishing a date when a technology will occur (such as when a specified capability has been achieved or a technological product is ready to enter the marketplace)

- Establishing a probable date for a technological development to occur.

Professor Martino in his textbook on technology forecasting states that "...when it is necessary to use expert opinion, Delphi is a good way of getting it." In most cases, we disagree. (He favorably contrasts Delphi with conventional committee meetings—and we certainly agree—but he never mentioned idea generation or the nominal group technique.) We rarely use the Delphi method, rarely recommend its use to our clients, and rarely see it used by corporations in their technology forecasting and planning. As mentioned above, Delphi usually takes too much time and money and the results are often inadequate. There can be no presumption of accuracy as a result of a consensus of experts. This fact was recognized by several of the Rand researchers, who realized the shortcomings of their own innovation and moved on to other forecasting tools, including cross-impact analysis and scenarios. Furthermore, there is a serious risk that the Delphi managers will build too much of their own bias into the way the questions are phrased and the manner in which the questionnaire is structured. The very structure that allows the participation of many experts inhibits the creativity of individual experts.

Yet, there are circumstances when the Delphi method is appropriate. One important one is corporate politics or marketing. Occasions may arise when dozens of people within the company or client base must be asked to participate and when interviews, one-time questionnaires, and group meetings are not adequate. Another is when the topic is so controversial that only an impersonal questionnaire and an iterative consensus would produce actionable results. Yet another is when you want to cultivate your clients or customers and let them participate in your study. In this situation, respondent satisfaction may be more important than the "accuracy" of the results, and the Delphi method is entirely appropriate.

Occasionally we hear of Delphi studies being done by corporations for technology forecasting and planning. Delphi is still used by some European and Japanese companies, but its use in the U.S. is declining. We expect less use of Delphi, and more use of questionnaires, in the 1990s than today.

Idea Generation. Battelle has been practicing idea generation techniques for the last two to three decades. Ours is a particular method, whereas other companies conduct "focus groups," "idea groups," "creative sessions," etc. Their purpose is roughly the same: to bring together a relatively small group of experts to generate thoughts on a defined problem for a stated goal. The applications are very similar to those of Delphi, specifically:

1   Identifying new applications for an existing technology or product

2   Identifying candidate technologies or products to solve a current need

3   Identifying issues and factors to be included in a larger forecasting method

4   Identifying implications and candidate strategies from a forecast as part of the planning process.

Typically, idea generation is not itself a forecasting tool because it usually will not produce a forecast per se. Rather, it is typically used as an input to a forecasting study or input to the planning process following a forecast. Also, it literally identifies ideas without evaluating them further. These ideas need further analysis to demonstrate feasibility.

In Battelle's idea generation method, the procedural steps are:

1   **Convene a group of 8 to 12 experts** and brief them on the topic and the procedure.

2   **Allow the experts to interact** by talking (brainstorming), writing down ideas (brainwriting), or some combination thereof. If the group is cooperative and creative, ideas can emerge from the group that would not have likely originated from any one participant acting alone. Meanwhile, the moderator records the ideas on large sheets of paper, which are taped to the walls of the meeting room for continuous review.

3   **Terminate the group interaction** when the participants show fatigue and a slowing down of discussion or at a predetermined time.

4   **Allow the participants to vote on 5 to 10 ideas** that they like best. This is usually done with markers or colored pens, with participants making their selections on the sheets of paper in the presence of the others. The open voting does to a certain degree encourage consensus, inasmuch as an individual may be influenced by how the others are voting. Any consensus may be convenient, but it is not required and not forced.

Idea generation is best used when the group is small, when the participants know and respect each other, when they are creative, and when the moderator maintains an even flow of interaction. It also works best on limited topics with little or no emotional or corporate political content. Furthermore, the participants must remain civil and not attack each other's ideas.

Conversely, idea generation should not be used in several circumstances: where more than 12 need to be included (one way around this limitation is to hold multiple sessions of different experts), when one person may try to dominate the group discussion, when the people do not know or do not like each other, when the topic is full of controversy and political content, and when the moderator anticipates difficulties in managing the interaction of the group.

The advantages of idea generation include creative spontaneity, the generation of rather long lists of potential ideas, and the good feelings of the participants if the session has gone well. The potential drawbacks include difficulties in group management (especially during brainstorming exchanges), the need to filter through the long list of potential ideas to evaluate them further for viability, and sometimes the lack of satisfaction by the participants that the session had been truly creative and responsive to the stated problem.

Idea generation for technology forecasting has some similarities to the focus groups practiced by market researchers. Focus groups bring together customers, or knowledgeable surrogates, as the "experts." The people are encouraged to react to a material product or a concept. Their behavior is a test of future consumer reaction to such a product or concept. In one sense, focus groups are experts on consumers; in another sense, they are themselves a test market. The fundamental difference between idea generation and focus groups is that the principal purpose of idea generation is to identify issues or suggest solutions to a problem, while the primary function of focus groups is to gather information. By design, the focus group has little structure to allow the fullest expression of opinions by the participants.

Nominal Group Technique. This technique for managing expert judgment in a group setting originated with Professors Andre L. Delbecq and Andrew H. Van De Ven at the University of Wisconsin-Madison in the late 1960s and early 1970s. It has a systems engineering rigor that particularly appeals to engineers. Whereas the Delphi method is on the wane in industrial applications, the nominal group technique is increasing in popularity as a group method, especially in technology companies.

The principal advantage of the nominal group technique is its structure, and its principal disadvantage is also its structure. So, it is best used when structure is needed, such as in the following circumstances: when certain people who can be argumentative and domineering must be included in the group, when people who do not know each other are together, when people who do not like each other are together, when managers and staff analysts are mixed, when the topic is sensitive or controversial, and when corporate politics need to be managed carefully so that the group exercise does not do more harm than good. When these circumstances are not present, idea generation may be more creative and produce richer results than the nominal group technique.

We use the nominal group technique in many of the same applications as those for idea generation. We have used it to identify factors to be included in scenario analysis. We have used it to identify candidate corporate strategies in response to scenarios. We have also used it to identify new applications for existing technologies. The nominal group technique can also be used successfully for generating criteria by which options can be screened toward a decision (which is very close to the original application by Delbecq and Van De Ven, who sought consensus criteria by which the federal government could meet welfare needs).

The procedure for this technique follows this order:

1 **A briefing is provided on the topic and the method** (usually by the moderator) to an assembled group, usually numbering 8–12 (although up to 18 can be accommodated).

2 **Ideas are silently generated on paper** by each participant. In idea generation, the group often talks out its ideas, but in the nominal group technique, each individual makes a list of his or her own ideas before the discussion begins.

3 **Each participant shares one idea from his or her list in turn,** in a round robin fashion. This allows each person to talk in turn and gives everyone an equal opportunity to share his or her ideas (which is a good way to get the quiet ones to speak and the talkative ones to be quiet). The moderator records each idea on large sheets of paper, which are mounted on the walls for continuous review (just as in idea generation). The round-robin should last three or four rounds, or until a specified period of time (a minimum of two hours) passes. During this step, participants are allowed to ask each other questions for the sake of clarification, but they are not allowed to debate or even comment on the virtues of other people's ideas. This rule limits creative spontaneity, but it also limits the risk of losing control that exists in the less structured idea generation technique.

4   The objective of a **review and consolidation of ideas** is to have the group review what it has generated and to see whether ideas can be consolidated to reduce redundancy. It also stimulates discussion on the meaning of the ideas. If someone objects to an idea being consolidated into another one, consolidation should not occur.

5   The **voting is done privately** on paper by the individuals. Many different variations are used. We pass out eight cards or pieces of paper to each individual, who then selects the eight ideas from the master list that he or she judges are the best ones (most germane to the topic) and records one per card/piece of paper. Then the individual goes back over his or her eight choices and assigns point values to each: eight points to the best, seven points to the second best, etc. (We often have to remind people that this exercise is "football" not "golf"—the high score "wins.")

6   The **tabulation of votes** can be done quickly in front of the group as soon as all votes have been turned in to the moderator. We list the ideas by number on a large sheet of paper and record the votes that each received as marked on the cards/pieces of paper. For example...

Idea 1.   2 pts, 4 pts, 3, 6, 8   [5 votes/23 pts...]

   2.   3, 4, 8, 7   [4 votes/22 pts]

   3.   (none)   [0/0]

   4.   5   [1/5]

If the group numbers eight, the most votes an idea could receive is eight and the most points is 64 (8 votes times 8 points). Typically, two or three ideas will receive more than six votes and 40 points, one or two ideas will receive four or five votes, and several ideas will receive only one or two votes. The moderator can go over the results of the voting so that all the participants learn the results before they leave the meeting. As with idea generation, a written report should be prepared on each session to recount the procedure and results (including tables with the master list of ideas and the record of the voting).

The voting measures the degree of consensus about the "best" ideas. Of course, these ideas have to be evaluated further according to their purpose, but the voting offers a valuable hierarchy of ideas to be considered.

Apparently, the nominal group technique is "nominal" in the sense that the collected individuals produce the ideas in a group environment, but the group does not literally produce ideas collectively—as in idea generation. The consensus, if one emerges, is not in the generation of ideas, but in the vote/point count for them.

Some common guidelines should be followed for both idea generation and nominal group technique sessions are:

1   Select the participants carefully. You need people who have familiarity and experience with the topic, but they certainly do not have to be the preeminent authorities on the subject. Often the "best" experts do not contribute the "best" performance in a group session. You also need people who are reliable: they will show up and will contribute according to instruction (the letter of invitation and the briefing paper). Furthermore, there may be corporate political reasons why some people must be included. We strongly recommend, when using group dynamics in forecasting, conducting at least two different sessions: one of company people (the microscopic expertise and corporate buy-in to the subsequent results) and one of outside experts (for the macroscopic perspective without corporate bias).

2   The number of people should be limited to eight to twelve. With the nominal group technique, you can use as many as 18 people, although we do not recommend that many. You should hold more sessions if you want to include more people. We have held as many as five different sessions, with five different groups of experts, in five different cities on the same topic. One group can never "validate" the results of another, but they certainly do supplement results for greater comprehension.

3   Preparation should always be extensive. You should select your participants carefully. You should write them a letter or memo of invitation stating the purpose, place, and time for the group session. You should also attach a one- or two-page briefing paper with the topic sentence, definitions, and outline of the procedure.

4   The meeting of the group should be away from the normal work place of the participants. At the least, it should be

in a room free of the telephone and other distractions. Concentration and dedication to the task are vital for success.

5   The tone, which is set by the moderator, should be friendly but businesslike. The procedure should be fun, and often is, but it is still serious business requiring serious effort.

6   Refreshments should be available, and at least one break should be taken.

7   Usually, a half day is required for a thorough group dynamic session. Idea generation can be performed in two or three hours with fewer than eight participants. The nominal group technique requires three to four hours. The maximum time allowed should be one day. We have conducted one-day manager retreats where we have briefed scenarios and performed a nominal group session in the morning and an idea generation session in the afternoon. After one day, unless the problem is large and particularly important, the participants will soon get restless and contribute less meaningful ideas.

Group dynamic methods, especially idea generation and the nominal group technique, do not require extraordinary resources. The first requirement is an experienced moderator. Moderators need to know how to set the proper tone and how to manage the group. They do not need to be experts in the topic, although they do need to be familiar with relevant terms that may emerge in the session. They must not be a partisan of any particular view that will make them seem biased to the participants.

The second requirement is participants. If they are outsiders, you may have to pay for their time and expenses. Typically, we do not. You cannot pay government officials for their time, but you may have to pay any associated travel expenses. Usually you do not have to pay academics, but then you need to share the information with them by providing them a nonproprietary report. Company people also represent a cost, but it is an overhead or opportunity cost rather than an out-of-pocket expense.

When Battelle conducts group dynamic sessions for clients, we typically charge between $5,000 and $10,000 per session. This includes our experienced moderator, preparation and organization, conduct of the session, and the follow-up written report. The cost may include travel expenses and pay for some participants. Also, we have to pay for room use if we hold it for a client at some off-site location, although the client may pay for this separately.

Group dynamics need not be expensive, but they do require some resources. Generally, they require more resources than major questionnaire surveys, about the same as interviewing, and less than the Delphi method. The time and money invested should be proportional to the importance of the results and the decisions based on them.

Group dynamic methods, especially variations on idea generation and the nominal group technique, are becoming more widely practiced by American and European companies. They have always been very popular in Japanese companies, although the techniques are different. The Japanese use highly organized committee meetings and focus groups to reach consensus, typically for decision making rather than forecasting and analysis. To date, the Japanese have shown little interest in doing idea generation and the nominal group technique themselves (although they have hired American experts to produce results for them using group dynamics methods); however, we expect the Japanese to begin to use these methods for technical creativity and innovation in the near future.

Based on our experiences over the last five years and lessons learned from clients and colleagues, the use of group dynamics is becoming more widespread. Often such sessions are called focus groups or idea sessions, and they employ many techniques. The nominal group technique, however, is becoming more widely known and used as a systematic approach to continuously unsatisfactory unstructured techniques.

Group dynamics will likely be used more in the 1990s than today because it offers a blend of creativity and group participation. It will not likely be used for forecasting per se, but in combination with other methods. Group dynamics are particularly well suited for issue identification at the early stage of forecasting and as option identification at the end of the forecasting and beginning of the planning stages.

Conclusions

Expert judgment has always been a major method or component of forecasting and strategy analysis. It will continue to be important in the 1990s. The key elements are whose judgment (a function of the experts), how broad-based the judgment (how many experts), and the reasons behind the judgment (the expertise). No one should ever accept expert judgment without explanation. Expert judgment can be very educational, and the information may be more important to the forecasting process than the accuracy of the forecast. Our expert judgment is that expert judgment alone is not a very satisfactory forecasting method, but is an excellent method of gaining information for use with other methods.

MULTI-OPTION ANALYSES

Multi-option analyses offer significantly different approaches to technology forecasting and planning than either trend analyses or expert judgment. In the other categories of methods, the underlying philosophy is that there will be one and only one future and that future can be known, or at least approximated, before it occurs. The methods of multi-option analyses, however, have a very different conceptual foundation: there may be only one future that will in fact occur, but we can never know with enough certainty what that future will be. So the approach is to estimate likely alternative outcomes for the future and plan toward at least one but, better yet, several of them. The objective is to remain flexible to the uncertainties of the future by recognizing the possible variations and anticipating responses to future circumstances as they come to pass.

The methods of multi-option analyses, not surprisingly, are typically used by companies and organizations that experience repeated and significant changes in their operating environments. Trend analyses, for example, may serve well companies with stable customers, market conditions, and business procedures. However, trend analyses can fail to meet the planning needs of companies where customers and prices are volatile or where market structure changes (due to politics and regulation/deregulation, nationalization or privatization, cartels, market confederations, etc.) In these highly dynamic conditions, companies turn to multi-option analyses to at least supplement, if not replace, traditional trend analyses. We have seen this shift during the 1980s in oil companies, electric and gas utilities, banking and finance, real estate, and airlines, to name a few industries.

There is another very important philosophical difference between multi-option analyses and trend analyses and expert judgment. Not only is the future uncertain, but it is also influenced in part by what we ourselves do to make it. In other words, there is often an implicit assumption in trend analyses and expert judgment that the future will be the future no matter what we do—it is all rather beyond our own control. (This attitude seems to come directly from both the ancient Greek and Roman beliefs in "fate" and the Judeo-Christian doctrine of divine determinism and predestination.) In multi-option analyses, the focus of inquiry is not so much on the question "what will the future be?," but rather on "what are the likely outcomes in the future, what is the outcome that I would most like to see happen, and under what conditions would my desired outcome really occur?" From the consideration of alternative outcomes, originate the forecasts; and from the forecasts, come the strategies that are most likely to produce the desired results.

In fact, many of the multi-option analyses techniques are not strictly forecasting methods, or at least not in the eyes of quantitative forecasters. Many statisticians, econometricians, and operation analysts do not recognize scenarios, trees, and portfolio analysis as forecasting methods. Shell International got around this barrier by calling its scenario approach "foretelling" rather than "forecasting." Whatever they may be called, the methods of multi-option analyses are certainly strategic planning, technology forecasting, and strategy analysis methods; and they may be predictive to the extent that we really do have an influence on our own future. We have worked with some companies that are largely reactive to their environments and can do little to change them. We have also worked with companies that were giants in their industry and their actions could have a profound influence on the future of their business. In most cases, we have concluded that companies have more influence on their future than they themselves believe or perceive.

Like the methods of expert judgment, the methods of multi-option analyses are not unique to technology. They, too, are general, domain-free methods borrowed from other disciplines and applied to technology. For forecasting technology in a narrow sense, these multi-option methods are not particularly effective. However, they are excellent for relating technologies with nontechnical factors.

Even if they have limited applications for technology forecasting, these tools are perhaps the best ones for macroscopic "product forecasting."

Broadly defined, multi-option analyses are the various methods that share the same fundamental approaches to forecasting and planning, including the consideration of several alternative future conditions and a desire to plan proactively for one or more of these alternatives. For this reason, Professor Joseph Martino, in his textbook on technology forecasting, characterizes this category of tools as "normative forecasting." Multi-option analyses are not normative in the sense of being so biased as to assert that "the" future will be what one wants it to be, but they are normative to the degree that they attempt to identify actions that will cause a desired future to occur (or be more likely to occur).

Four types of multi-option analyses are discussed below:

- Scenarios
- Simulations
- Paths and Trees
- Portfolio Analysis.

Scenarios

The concept of scenarios comes from the theater. Scenarios are a writer's outline of a plot. They contain the outline from which a script is derived. In their original context, scenarios provided a planning tool that was predictive only to the extent that the final script reflected the points in the conceptual scenario. One of the most famous examples was the process used by Walt Disney to create *Snow White,* the first full-length animated movie. Several key sketches were mounted on the wall. They were spread out to allow space for sequence sketches that fleshed out the major parts of the story. By the time the script had been thought out, sketches covered the wall. These sketches then provided the basis for thousands of filler drawings to achieve the effect of animation. Today, if you were to discuss scenarios with a Hollywood producer, he would assume you were discussing the outline of a script plot.

The first application of scenarios for planning outside of drama occurred at the Rand Corporation, in the 1950s and 1960s, when Herman Kahn became the guru of scenarios for U.S. Air Force strategic planning. In planning what it would have to do to wage war successfully with the Soviet Union if that catastrophe were ever to occur, the Air Force largely duplicated the strategic bombing approach of World War II against Germany and Japan, except for the extensive use of nuclear bombs. Kahn used hypothetical, cause-and-effect sequential scenarios (much like the script of a story) to demonstrate to the Air Force the need to consider alternative approaches. In this application, scenarios were also planning techniques, as with scripts, but with alternative paths and alternative outcomes. Kahn did not place probabilities on his scenarios and he did not use them as predictive forecasts, but he did use them to examine the conditions that would need to exist to expect certain consequent outcomes. Also, in this context, the scenario that the Air Force implemented, if necessary, would greatly determine the resultant future circumstances of the globe. Today, if you were to use the term "scenario," an Air Force officer would assume you were discussing strategic plans.

The first principal industrial application of scenarios was by General Electric (GE) in the late 1960s and early 1970s to anticipate general American social and economic conditions by 1980. The GE analysts shifted the use of scenarios from different paths with different outcomes to just alternative outcomes. The emphasis was on a thorough description of alternative future conditions by the year 1980 without trying to guess the actual path to that year. In this application, the scenarios were primarily anticipatory, if not predictive, of the external market environment without being first a planning exercise (as with the applications by Walt Disney and Herman Kahn).

Perhaps the most famous corporate application of scenarios is that of Shell International (and its sub-corporations, Shell UK, Shell Nederland, and Shell Canada). Shell built on the GE experience to anticipate different prices for a barrel of oil and to plan the company's possible responses. Shell's scenarios were predictive to the extent that one or more of them did generally capture a future condition. Shell did not predict the occurrence of the OPEC oil boycott and the subsequent quadrupling of oil prices, but it did generate a scenario with an oil crisis that Shell managers considered before it really happened. More importantly, Shell's scenarios were excellent learning tools for contingency planning. In a very unstable environment of fluctuating oil prices, Shell management adapted remarkably well and produced profits when other oil companies did not.

In addition to the different types of scenarios, there are many methods for generating them. A major source of confusion about scenarios is due precisely to the fact that there are various types, methods, and applications. There is no single or even "right" method of using scenarios, but rather there is a cluster of techniques.

Some methods are highly imaginative and qualitative, somewhat in the flavor of Disney and Kahn. Indeed, these scenarios are often referred to as "intuitive." The refined method of Shell and SRI International is highly dependent upon the group that generates them, the specifics of oil and gas (domain specific), and the corporate culture in which the scenarios are reviewed and planned upon. This approach uses trend analyses and expert judgment extensively as inputs to the scenarios. The analysts focus on one or two critically important issues and hypothesize two, three, or four likely outcomes. Then they flesh out each outcome with an elaborate story incorporating other issues associated with the different outcomes. These scenarios become archetypical because each scenario is very different from the others, as though if some critical factors varied, then all other factors had to vary in their outcomes as well. They are, therefore, rarely predictive, because what does occur is often a shade of gray between two scenarios. However, they provide guidelines for monitoring events and responding to them with contingency plans.

Beyond Shell, the qualitative, intuitive scenario method has been applied by such diverse companies as British Petroleum, BellSouth, Pacific Bell, Florida Power & Light, and Southern California Edison, to name but a very few.

Another approach to scenario generation employs the techniques of cross-impact analysis, which was developed in the 1960s by analysts at the Rand Corporation (several of whom had been involved in the development of the Delphi method and scenarios). This approach involves a matrix of factors (or trends or issues— variables, in general) and asks how one factor impacts the other factors. For an example of a cross-impact matrix see Figure 9. Cross-impact analysis seeks to integrate all the trends to find the net solution. Because we cannot be certain that one solution alone is possible, the result of cross-impact analysis is a distribution of solutions or scenarios. The different scenarios may contain similar outcomes for some factors, but different outcomes for others. (Theoretically, this approach asserts that the future will be a hybrid of some continuing trends and some discontinuities, but not altogether different from today or the past.)

The mechanics of cross-impact analysis also vary widely. One technique is rather qualitative and typically considers one outcome for each factor (trend impact analysis as refined by Ted Gordon of Rand and practiced by The Futures Group). Another technique is highly quantitative and uses statistical simulation along with scenario analysis (the type of cross-impact analysis refined by Selwyn Enzer, a protege of Olaf Helmer of Rand, and practiced by the business school of the University of Southern California).

Yet another technique was developed by Battelle in its scenario method, which is called BASICS. It has been used by Rockwell International, Goodyear Aerospace, Los Angeles Department of Water and Power, The Gas Research Institute, BellSouth, Lucas Engineering, General Motors, and many others. This approach provides multiple possible outcomes for each factor and cross impacts them as conditional probabilities (as expressed by index values rather than as probabilities) upon judgmental marginal probabilities (which are expressed as probabilities). The algorithm that calculates the probability adjustments is deterministic. It does not employ Monte Carlo statistical probability routines as found in the Helmer-Enzer approach. (The deterministic algorithm may be unsatisfactory for those who seek forecasting accuracy based on statistical sophistication, but it is indispensable for simulation. Answers vary only according to inputs and not also through the process of calculation.)

Development D_i	Probability P_i
D_1 One-Month Reliable Weather Forecasts	0.4
D_2 Feasibility of Limited Weather Control	0.2
D_3 General Biochemical Immunization	0.5
D_4 Elimination of Crop Damage from Adverse Weather	0.5

If This Development Were to Occur:	Then The Probability of			
	D_1	D_2	D_3	D_4
D_1		0	0	↑
D_2	↑		0	↑
D_3	0	0		0
D_4	0	0	↓	

FIGURE 9. EXAMPLE CROSS-IMPACT MATRIX

All the different definitions and methods of scenarios share certain common features, such as

1   Rational, explicit, and documented analysis of trends and events by a target year (sometimes expressed as probabilities, even if only as a hierarchy of likelihood, sometimes not)

2   Alternative sets of factor outcomes (scenarios provide multiple outcomes that reflect uncertainty in the future rather than a single "must be" prediction)

3   Sets of outcomes that are internally consistent with each other, so that scenarios are logically unified internally and distinctly different from each other

4   Views of the future (or futures) that provide a foundation for contingency planning and decision making (including elements of communication, corporate politics, and corporate culture).

Although there are different types and definitions of scenarios, the definition that Battelle uses for BASICS is typical of others as used in the corporate environment in the 1980s. Scenarios are descriptions of alternative, but internally consistent, sets of factor outcomes derived from a logically disciplined method for the purpose of forecasting (or foretelling) and strategy simulation (or evaluation).

Like all the other forecasting methods discussed in this report, scenarios are well suited for certain types of applications and not well suited for others. In general, scenarios are most effective when these demands are present:

1   **Macroscopic factors** beyond the scope of quantified variables, and established relationships among them, **need to be included** (or, as Professor Martino observed, when nontechnical factors, such as politics and economics, are very important relative to the technical factors.) Scenarios are excellent at integrating political, social, economic, and technological factors at the most aggregate level of analysis. Because of its potency at the macro level, scenarios are difficult to use for precise, micro answers, such as those produced by econometric and statistical models.

2   **Long-term time frames are required.** It has been well demonstrated that most types of trend analyses, particularly the quantitative ones, lose much of their accuracy and utility the farther out into time they have to predict. Many of the quantitative trend analyses methods are not reliable beyond one, two, or three years. They certainly are inadequate out to 10 years or more, especially in unstable conditions (such as with oil prices and all forms of energy supply that are dependent upon many different and dynamic factors). Scenarios are estimated, general portrayals of future conditions and are well suited for addressing great uncertainty.

3   **Static descriptions of future environments are useful.** As a planning tool, scenarios are sequential by step, if not by time. As a forecasting tool, scenarios typically are neither step or time sequential. Scenarios have been used largely to capture a target year, but do not attempt to capture exact sequences of events by exact time increments. This approach is the legacy of the GE scenarios as they differed from Kahn's scenarios. If precise points are needed by time increments, then one should use trend analyses, particularly the quantified ones, rather than scenarios. Better yet, one could use scenarios as alternative points (or estimated clusters of points) for the future, with today as another point, and then use the quantitative methods to interpolate the points in between.

4   **Great uncertainty surrounds the problem.** Scenarios do not ignore uncertainty, but rather deal with it directly by using alternative outcomes and judgmental probabilities. Shell has demonstrated that a company can use scenarios to think through the uncertainties of one's business environment. The benefit of this approach is "what do we do now?" can be considered before one faces a situation rather than after. As the old saying goes, to be forewarned is to be forearmed. Scenarios at least give you the opportunity to be proactive rather than reactive to your business circumstances.

5   **Data are scarce, unavailable, or very expensive.** Many quantitative methods are data intensive and data expensive. Scenarios are not as data intensive, although they can be judgment intensive.

6   **Non-quantitative factors must be included** in the analysis. Too often quantitative methods have simply ignored factors when they could not be quantified. These ignored factors have included such things as OPEC political goals, Soviet political objectives and military doctrine, social attitudes in the U.S. concerning nuclear reactors, courts breaking up corporations, governments privatizing corporations, etc. Scenarios allow qualitative analysis as well as quantitative analysis. In the former, the burden is upon definition rather than measurement.

Scenarios have been best used in corporate planning by providing estimates of future external environments, including future demand for products and marketplace conditions (economic, social, and political). They can also be used as simulation models. The general way to do this is to first generate scenarios of the future topic with the best available information and skills. What will the scenarios be if all factors and trends do play out as estimated? Once that question is addressed, one can ask what can be done to make the scenarios we would like to see happen more likely to occur? Then one can start varying inputs to test outcomes (as long as the analytic process or algorithm is kept constant). Our expectation is that in the next one to two decades, as a simulation method, scenarios will be much more widely used than they are now (which is not very).

Scenarios as practiced in the 1970s and 1980s have tended to be manpower intensive, time consuming, expensive, and too often too difficult to communicate and disseminate within the company. The Shell and SRI method is particularly expert intensive. Often scenarios take a year or more to generate at a cost of $200,000 to $500,000. The Futures Group, University of Southern California, and Battelle have attempted to package and sell their methods (as supported by personal computer software) to lower total time and expense. Scenario analysis, like other methods, needs to be flexible enough to provide "quick and dirty" answers, as well as highly refined ones, according to the needs of upper level managers. We have performed a BASICS scenario analysis on a very discrete question within a day using two analysts and our own method and software. We were not comfortable with the lack of precision in our analysis, but we were responsive to the time-urgent needs of our boss's boss's boss.

Another drawback of scenarios is that they tend to be broad and conceptual rather than specific. Their generality makes them difficult to use for planning by many middle-level managers. Also, managers often require numbers in their plans, and they may not get as precise a set of numbers from the scenarios as they are used to receiving from trend analyses. One solution to this problem, obviously, is to include numbers, even as ranges, in the scenarios and use some expert judgment to make the transition from scenarios to trend analyses. Another solution is to use the scenarios as input to micro models that do produce the numbers required for managers' decisions.

Scenario analysis will likely increase in popularity in the 1990s. It will not likely be used alone, but rather in conjunction with several other methods and techniques (both trend analyses and expert judgment). For macroscopic and long range forecasting and planning problems, scenarios will become the predominant method because, in theory and mechanics, scenarios are the best method.

Simulations

Like scenarios, simulations are various approaches that are clustered together around a common theme. Simulations are elaborate games for the purpose of understanding and testing rather than entertainment. Popular games such as chess and pachisi (Parcheesi) were developed centuries ago as simulations of warfare. Their value for learning the rules of war seem long faded, but they provided examples for much more elaborate war games in the nuclear age.

All simulations, and they vary as greatly as games, have two common elements. One is the model of a past, present, or future situation. This model is a representation of a more complex reality. In a very simple sense, a game board, equipment, and rules provide a model (such as Monopoly, which is a simplified and humorized model of Atlantic City, New Jersey, c. 1930). The model might also be a set of very complicated mathematical equations that can be handled only with a computer. The second element is the "playing," or the repetitive routines that demonstrate that changes in inputs produce changes in outputs (as long as the model itself remains constant). Therefore, simulations can provide a safe, analytical learning and testing experience that cannot be directly replicated in reality, or when being "right" or "wrong" may be disastrous. For example, pilots learn to fly in flight simulators before they learn to fly in a real aircraft, where accidents are often fatal. Another example was the NASA flight simulator for the first manned landing on the moon in 1969. In numerous landing simulations, pilot Neil

Armstrong "crashed" into the simulated surface of the moon because of miscalculations. When the real landing occurred, however, Armstrong made a perfect landing.

Perhaps no activity of man has been gamed or simulated as much as warfare. From simple chess to full field maneuvers, armies have continuously practiced their craft in preparation for the real thing. In this respect, gaming and simulations are a form of rehearsal, as though for a play or a concert. But they are more than rehearsals in the sense that they have a strong experimental content. In a rehearsal, the action is repeated until a desired and expected result is achieved. In simulation the action is repeated until an unknown procedure is identified for a desired result. Simulation asserts that you can influence your own future if only you know how to achieve it.

Simulation (the term we will use broadly to include both quantitative and qualitative techniques, including gaming) has not yet been widely practiced by corporations in their technology forecasting. It has been extensively used in training, but not by executives in decision-making. However, simulation techniques serve to remind us that once a model has been developed, it should not only be used to forecast the most likely future, but also should be used as a strategy analysis tool. After all, the purpose of studying the future is not merely to satisfy intellectual curiosity, but to determine what actions taken today might best affect the future in your favor.

The advantages of simulation are that:

1   They provide exciting and insightful learning about the modeled reality and how to manipulate it successfully ("winning"). People typically enjoy games, and they enjoy business simulations, too. Simulations provide an intellectual challenge with no real risks, and they develop empathy toward the problem through direct personal experience. In addition, simulations encourage interpersonal communication and cooperation through the shared experience.

2   They provide a relatively risk-free opportunity to experiment with a situation that in reality can have very great risks. Simulations provide a means to test and explore. With a model of a situation, you can experiment as though you are in a laboratory: change this input and compare changes in results, etc. If a reality is, in fact, modeled accurately, simulation can be used as a forecasting tool as long

as your own actions do in fact shape the future and the model itself stays constant.

On the other hand, simulation methods have serious disadvantages, which explain why they have not been widely popular with upper level corporate managers. These shortcomings include the following:

1   Models often are not accurate or constant. Too often they are simplistic, if not simple-minded, and taken far too literally. Errors in modeling can lead to errors in simulating and consequently errors in decision making. In many cases, simulations require a level of analysis and critical judgment beyond the resources of companies where profits depend on short-term action more than long-range planning.

2   Models can be time consuming in construction (literally or figuratively on the computer) and very expensive. The major costs are in the manpower to research and create the models, especially highly complex "realistic" models. Furthermore, the more realistic the models, the more complex they are; until the simulations become so complex that they deter managers from using them. Managers do not enjoy using difficult decision-making methods to make difficult decisions.

3   As difficult and expensive as the models can be, they can also be difficult to update and maintain, particularly if the reality being modeled is dynamic.

These disadvantages, however, are not insurmountable. Personal computers are becoming more powerful and easier to use. Software will eventually catch up with hardware, and managers will have at their disposal some rather sophisticated models and programs. As these simulation programs become available, both as generic tools and as industry-specific models, simulation will become easier to perform, easier to communicate, and easier to apply.

Simulation is unlikely to become a highly accurate forecasting tool because too few players (even the U.S. Air Force) are really potent enough to determine their own futures. In some industrial situations, when a company enjoys a very large market share for a product, the actions it takes may in fact largely determine its future. In this situation, simulation becomes very valid and worth the resources if indeed the stakes are not insignificant.

Paths and Trees

Although planners often do feel as though they are lost in the woods, paths and trees are planning tools rather than wilderness characteristics. They are different names for the same approach: given a goal, how can one visualize the different ways of reaching it. The approach has been described as a segmentation technique. It has been called a "path" because, with it, a mental route can be traced through a labyrinth of options. It has also been called a "tree" or "relevance tree" because, when drawn, it looks somewhat like a tree sticking out of the ground with a large central trunk and many branches. It might have also been called a river with many tributaries, each with its own network of feeding streams. Paths and trees are used to structure a topic (see Figure 10) for analysis of a problem (see Figure 11) or a plan.

Trees and paths became very popular in the 1970s. Since that time, however, they have declined in use and appear to be declining further. Over the last two decades, virtually every major corporation has used paths and trees at some time to structure its thinking about strategy and the future.

Today, we know of no company currently using paths and trees for major decisions. The approach has several advantages, which account for its popularity over the last two decades. For example,

1   They do provide an excellent tool for structuring a problem, especially a planning problem for which the goal has been identified and the sequence of steps to achieve that goal need to be thought through. The nodes of the tree, or the forks in the path, can be either "and" or "or" options: at one node, we need to do two things simultaneously, or we must decide to go one way or another (and think through where that will eventually lead us).

2   They provide a means, like PERT charts, to plan a task through multiple stages toward completion. As a visualization, paths and trees can be communicated to other people to win their cooperation.

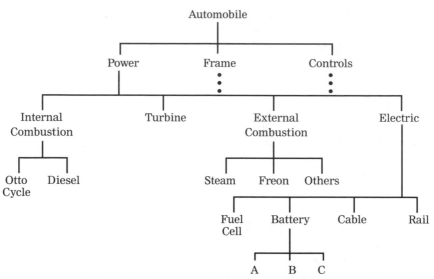

Source: Adapted from Martino, *Technological Forecasting for Decision-Making* (1983), p. 160. Reprinted with permission.

FIGURE 10.   EXAMPLE OF A RELEVANCE TREE
(Components of an Automobile)

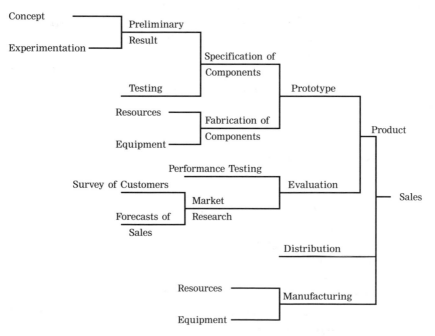

FIGURE 11.   EXAMPLE OF AN R&D PLANNING TREE

3   They are not particularly difficult to construct, do not require extraordinary manpower, and can be achieved rather quickly. In addition, they do not demand a great deal of data or complicated techniques. They are thought intensive, however, and very much dependent upon the mental skills of the analysts preparing them.

On the other hand, paths and trees suffer several shortcomings, which explain why they have declined in use. These problems include the following:

1   They can overstructure a problem and preclude consideration of alternative outcomes. They should be used only after a goal has been identified; they do not particularly help in evaluating alternative goals, unless several alternative paths and trees are considered.

2   They are not really forecasting tools, except in cases where plans can be successfully implemented. In rare situations, they have been used as tools forecasting emerging technologies. The best use of paths and trees is at the tactical, implementation stage rather than at the strategy and concept stage.

We have used trees in a technology forecasting problem (see Figure 12) We asked the question: how likely is it that a certain computer chip technology will become the prevailing market standard by the year 1995? To answer this question, and to assign a probability to it, we asked experts to identify the developments that would have to occur first for these types of chips to succeed. These input developments could be structured as a tree with branches leading to the trunk. Some had to do with manufacturing techniques. Some had to do with materials developments. Other developments concerned inspections, packaging, and testing. We assigned judgmental probabilities to the occurrence of each individual development and then mathematically determined the net probability of final computer chip development and use. We answered the question posed, but we ignored other possible answers, such as a forecast of competing chip technologies.

The Futures Group has recently developed a new approach to trees and paths as a technology planning tool. They call their approach "forecasting backwards." The approach begins with a desired, hypothetical technological parameter. What options might be available for achieving this parameter? With this question, analysts construct option trees stemming from the desired result. Research and operational paths can be identified and simulated for length of time, cost, and likelihood of success. This approach, like scenarios, stresses "normative forecasting" of technology, along with many other business factors, toward a comprehensive "product forecasting."

Our expectation for the 1990s is that the use of paths and trees will continue to decline in popularity as a method for technology forecasting. They may well increase in use as a planning tool for R&D in the laboratories at the tactical level of implementation, which appears to be the most valid application for paths and trees.

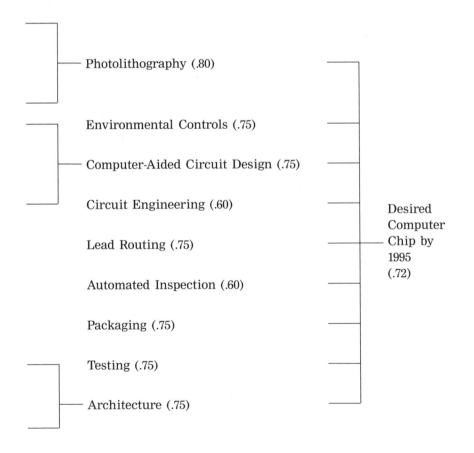

FIGURE 12.   EXAMPLE OF A TECHNOLOGY FORECASTING TREE

Portfolio Analysis

Portfolio analysis is a method of financial risk analysis that has been applied to technology planning. It is not a forecasting method, because it offers no foresight about the future development or performance of technologies. It is a planning tool in a very general sense, but not a method that identifies actions. Most of all, it is a way to assess the business risk (investment vs. payback) of new product development.

The best known variation of portfolio analysis is the risk matrix of the Boston Consulting Group (BCG). It is essentially a quadrant of High/High (positive/positive), High/Low (positive/negative), Low/High (negative/positive), and Low/Low (negative, negative). An example appears as Figure 13. The labeling of the axes varies greatly, depending upon the specific application. In the original BCG matrix, the horizontal axis is Relative Market Share (as a measure of cash generation) and the vertical axis is Growth (as a measure of cash use). But the two axes can also be Cash Return and Cash Invested, Market Success and Resources Required, or Technological Development and Financial Risk. In all these cases, the matrix cannot predict technological development or market success. But the matrix does categorize the technologies or products into business risk groups for business analysis and decision making. Variations in the portfolio analysis matrix are given in Figure 14.

The principal advantage of portfolio analysis is the visualization of categories of risk that technologies represent as corporate resources. For upper level managers who rely heavily on visualization of ideas, portfolio analysis provides an effective communication tool. It also appeals to managers with financial backgrounds. As the name implies, this type of analysis is borrowed directly from investors who use it to evaluate the performance of their stocks and bonds, and those they wish to buy and sell. Unfortunately, evaluating technologies and market potential in the future is just as difficult as evaluating the performance of existing stocks. Portfolio analysis suffers from several disadvantages as a forecasting and planning method. As shown above, there is so much flexibility in labeling the axes that portfolio analysis can be applied in almost any situation and about any answer can be derived. In other words, the use of the method is not only highly judgmental, but also highly arbitrary (even biased). Too often this method is used for ex post facto rationalization of a favored policy rather than as an ex ante tool of analysis. Beyond the issue of labeling axes, to continue this same

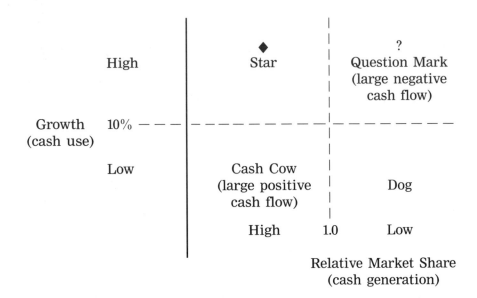

Source: Michael E. Porter, *Competitive Strategy* (1980), p. 362. Reprinted with permission.

FIGURE 13. PORTFOLIO ANALYSIS MATRIX

A. Base Matrix

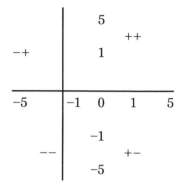

B. Cash Flow Matrix

FIGURE 14.   BATTELLE VARIATIONS ON THE PORTFOLIO ANALYSIS MATRIX

C. Technology Risk Matrix

	Emerging	Mature
High	High Risk but maybe cash cow of the future	Hopefully, backbone of business
Low	Safe Bet	Declining Products and Services (cash cow)

Financial Risk (Level of New Investment)

Technological Development

D. Tech Risk Array Matrix

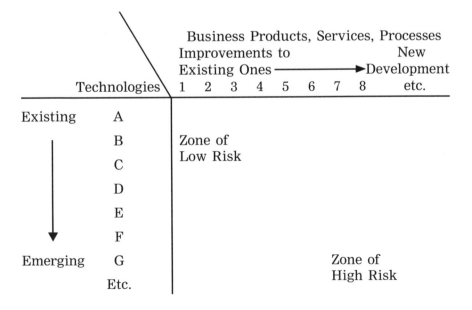

Business Products, Services, Processes
Improvements to   New
Existing Ones ⟶ Development

Technologies   1   2   3   4   5   6   7   8   etc.

Existing   A

            B   Zone of
            C   Low Risk

            D

            E

            F

Emerging   G   Zone of
                                            High Risk
        Etc.

FIGURE 14.   (Continued)

point, calibration of the axes is often vague. Therefore, placing technologies in the matrix can be very arbitrary.

A second problem is that portfolio analysis tends to become rather simplistic and even misleading when trying to assess unproven technologies and untested new products. The method works much better when applied backwards with existing data to evaluate existing products, for either further investment or for divestiture. It does not work well looking forward with so many unknowns.

A third problem is conceptual: portfolio analysis tends to stress short-term and fragmented returns. It is not a tool of long-range investment (and not widely used by the widows and orphans of the stock market) and of technology synergy. On the broadest scale, portfolio management as a way of evaluating corporate divisions and business units has been vigorously attacked by management experts, including Michael Porter of Harvard, as being short-sighted and uncreative.

Although portfolio analysis has its shortcomings, it can have some practical benefits for technology forecasting and strategy analysis when used with some other methods. One application has been used by Shell. Shell took several of its archetypical scenarios and arrayed them as columns (as though they were a horizontal axis) and took candidate strategies and arrayed them as rows (as though they were a vertical axis). By using expert judgment through consensus, they filled the matrix with weighted scores to determine the strategies that scored best relative to each scenario. Through this process, Shell was able to prepare contingency plans (strategies) for each particular scenario. This approach could be applied to technologies by arraying the candidate technologies in place of strategies to determine expected winners relative to each scenario (which need to include such factors as politics, regulation, consumer demand, etc.).

We have tried another variation with several middle managers by first deriving business implications from one scenario using the nominal group technique. We set up a matrix of these implications and current products to evaluate how each implication might be good or bad for product growth. From this exercise, the managers gained insight into the level of risk and opportunity their products faced in the future.

Another adaptation of portfolio analysis for technology planning is called Tech Risk Array, which has been developed by Battelle (See Figure 14, part D). It also, like the Shell approach, breaks away from the confined quadrant. On the far left side of a matrix (the rows) a range of existing candidate technologies to highly speculative possibilities are listed. On the top of the matrix, as column headings, business products are listed—from improvements to existing services and processes (on the left) to very new developments (going out to the right). The cells of the matrix contain words or numbers reflecting the evaluation of risk. The upper left corner represents little risk, while the lower right reflects great risk. This array can be combined with scenarios, much as with the Shell approach, and with the BCG quadrangle to provide perspectives on the amount of business risk associated with certain technologies.

To repeat, portfolio analysis is a method of assessing the business risk and potential of a technology as goods and services. It provides a useful tool for anticipating the financial aspects of technologies, but it cannot predict technological performance or timeframe. As it exists today, we believe portfolio analysis will decline as a method of technology planning because of its conceptual limitations and nontechnical orientation. However, hybrids of the method, when used with other methods (especially trend analyses and scenarios) may present new analytical possibilities in the 1990s.

Conclusions

In most applications, multi-option analyses are planning rather than forecasting methods. To the extent that plans do work and to the extent that it is possible to make one's own future, these methods indeed produce "normative forecasts." These approaches emphasize multiple outcomes (as forecasting) and multiple options toward a goal (as planning). They are particularly valuable for dealing with long-term horizons, uncertainty, and contingency planning.

C H A P T E R 5

CONCLUSIONS

Our goal in preparing this guide was to identify different technology forecasting and strategy analysis methods for the benefit of corporate managers. We have emphasized the advantages and disadvantages of the methods, when to apply them, and how to evaluate them. We have touched on the mechanics of the methods, but we have left the details to further research by the analysts.

General Observations

We made the following generalizations based upon this report:

1   Too much emphasis has been placed on the accuracy of forecasts and not enough on the educational and communication value of the forecasting process. The process should make both managers and analysts better informed about the topic and better prepared to deal with uncertainty. The ultimate objective of technological forecasting and strategy analysis methods, whether qualitative or quantitative, by whatever method is used, is to make experts out of managers and analysts who can exercise their own expert judgment in the decision-making process.

2   Goals and purposes must be identified for the forecast and plan before selecting the appropriate methods to achieve

them. The very first question is "What is the question?" The second question is "What do you intend to do with the answer?" Then the methods are selected and applied accordingly. A further understanding of the context of R&D, as outlined briefly in the Introduction, also provides guidance to the use of methods.

3   Methods must be used in combinations. No one method can answer all questions. Trend analyses, expert judgment, and multi-option analyses can be combined according to the needs of the forecasting effort. The extent to which these methods are combined is very much dependent upon the skills of the analysts and managers and the corporate climate.

4   Technology forecasting methods also need to grow to incorporate applicable features of economic forecasting, political analysis, and market research. In the broadest sense, technologies, and especially their final products, are profoundly affected by nontechnical factors (including economic, political, public policy and regulation, social preference, etc.). We need to grow beyond "technology forecasting," as narrowly defined and practiced, into "product forecasting."

5   Technology forecasting should accomplish three things:

—Provide an indication of the future technological environment and technology capabilities that the firm will either have or need to acquire;

—Suggest alternative technology strategies to the managers; and

—Evaluate these strategies to see which will likely produce desired results.

Thus, technology forecasting should be an integral part of the overall corporate strategy-making system. A good corporate strategic plan will encompass all the firm's intellectual property and technological assets and manage them for the best long-term return.

6   Virtually no method of technology forecasting and strategy analysis that we have examined is unique to technology.

Indeed, they have all originated in economics, finance, marketing, operations research, political science, mathematics, statistics, and the natural sciences (especially physics). The only methods that come close to being "unique" to technology are S-curves and patent trend analysis (and they are specialized variations of growth curves and trend analyses). On the other hand, not all forecasting tools are useful for technology forecasting. As just one example, the VALS system of SRI International is applicable only to marketing, not technology forecasting.

7   Battelle has seen recent growth in the number of specialists within companies assigned to perform the specific job of technology forecasting. Generally, this duty is performed in a separate department, with one to three professionals. They usually report either to the CEO's office, the director of R&D, or a similar senior manager. They often are funded at a relatively low level, but still have enough responsibility and authority to make technology investment decisions. They are funded on a relatively long-term basis, almost always more than one year, to allow time for technology plans to be implemented and evaluated.

Our expectation of the extent to which each technique reviewed in this guide will be used during the 1990s is summarized in Table 1.

A comparison of the techniques reviewed in this report is provided in Table 2. We offer our judgments of the principal technology forecasting applications, advantages, disadvantages, and resources required to apply each technique. "Manager's Guide to Forecasting," by David Georgoff and Robert Murdick (found in *Harvard Business Review,* January–February, 1986) provides an independent comparative appraisal of the applications and resource requirements for many of these methods.

Forecasts for the 1990s

The following are our expert judgment forecasts for the 1990s.

1   Technology forecasting and strategy analysis methods will become better understood and applied through trial and error in the corporate environment. The methods will become more sophisticated. We hope that they also become more responsive to the needs of managers. Meeting this challenge is the responsibility of both managers and analysts.

TABLE 1. CURRENT AND EXPECTED USE OF TECHNOLOGY FORECASTING TECHNIQUES

Method	Current Use For Specific Technology Forecast	Current Use For Several Technology Forecast Uses	Use During 1990s For Specific Technology Forecast	Use During 1990s For Several Technology Forecast Uses
Trend Extrapolation		3		3
Time Series Estimation		2		3
Regression Analysis		2		3
Econometrics	2		1	
Systems Dynamics	1		0	
S-Curves		1		1
Historical Analogies	1	1		2
Input-Output Matrices			1	
Patent Trend Analysis	1	1		2
Scientific Literature Analysis			1	
User Created Database Analysis		0		2
Interviews		2		2
Questionnaires		2		3
Delphi		1		0
Idea Generation		1		1
Nominal Group Technique		1		2
Scenarios	1			2
Simulations		1		2
Paths and Trees	2		1	
Portfolio Analysis	2		1	

Numbers used in this table mean:
0 = little or no use of technique
1 = used by a few companies
2 = used by many companies
3 = used by almost every company.

TABLE 2. COMPARISON OF TECHNOLOGY FORECASTING TECHNIQUES

	Technology Forecasting Applications	Advantages	Disadvantages	Required Resources*
Trend Extrapolation	Technology capabilities, product sales, time to develop new product	Inexpensive, easy to understand, lots of software available	Inaccuracy, not based on theory of causation, only short-term	Historical data over time, graphic paper or software, 5 minutes, $50
Time Series Estimation	Technology capabilities, product sales, compare rate of change in use of technologies	Inexpensive, more accurate than extrapolation	No theory of causation, resulting in inaccuracy, only short-term	Time series data, analysis software, staff that knows techniques, 1 hour, $100
Regression Analysis	Technology performance, product sales, development time	Relatively accurate for the expense	Must know causal variables and predict future values for them	Data on several variables, analysis software, 4 hours, $300
Econometrics	Longer term technology performance and product sales, time to develop new technologies	Equations keep forecast within bounds, causal in nature	May be too complex for accuracy achieved, especially long-term	Data on many variables, analysis software, 2-4 weeks, $8,000
Systems Dynamics	Determining what factors interact in technology development, inputs to scenarios	Assists in thinking through a problem	Complex, not well known, no generalized software, world not always cyclical	Staff skilled in method, some data collection, programmer, 1 month, $15,000
S-Curves	Market share, technology capabilities, contingency planning	Emphasizes limits to growth, inexpensive	Relevant technological data seldom available	Knowledge of related technology, 4 hours, $300
Historical Analogies	How much R&D to spend, type of research staff structure to use, identify implications of strategies	Avoids repeating costly mistakes, insight into how others solve problems	Only work when relevant data is available, which can be expensive	Case study or PIMS data, a problem may take one day to investigate, $100,000/yr
Input-Output Matrices	Effect of change in technology or process on economy	Computes total economic effect, including any "multiplier" effect	Requires I/O table, can be expensive for narrow solution to niche problem	Knowledge of I/O math, table of transactions, software, 1 month, $12,000
Patent Trend Analysis	Technology monitoring, investment decisions, identify new competitors	Insight into current R&D of competitors, software available to help	Higher cost than trends, data is 18-36 months old, not all R&D is patented	Some knowledge of method, access to patent database and software, 2 months, $25,000
Scientific Literature Analysis	Technology monitoring, company R&D analysis, identify developing areas	Literature more current than patents, software may be leased to help	Only a fraction of R&D work is published, search may not find relevant work	Knowledge of technique or consultant, literature database, 2-3 months, $50,000/yr
User Created Database Analysis	Technology monitoring, scanning, and planning	Works with any printed material	Not yet available, can get buried in irrelevant data	Scanner, computer, software programs, ongoing during year, maybe $50,000/yr

TABLE 2. COMPARISON OF TECHNOLOGY FORECASTING TECHNIQUES (Cont.)

	Technology Forecasting Applications	Advantages	Disadvantages	Required Resources*
Interviews	New product or process introduction, technology possibilities, trends	Requires no database, offers qualitative insights	More expensive than questionnaires, can take much effort	Interviewing skills helpful, 1 hour to 1 day (if travel), $100 per interview
Questionnaires	New product or process introduction, technology monitoring/performance, future technological environment, measure product demand	Access to many responses quickly and inexpensively	Follow-up difficult, structure to questions limits responses	Staff knowledgeable in design, software for tabulation, 30 minutes per instrument, as little as $25 per instrument
Delphi	New product or process introduction, identify key technologies for company (for large groups)	Forced consensus	Forced consensus, relatively expensive	Staff knowledgeable in Delphi and survey design, software to assist with tabulations, 3 hours per person, $300/person
Idea Generation	Identify new products, technologies (for groups of 8 to 12)	Creative viewpoints collected, little filtering of ideas	Very possible for one person to dominate, ideas may get suppressed	Knowledge of technique, good group of experts, 3 hours per topic, $5,000 per session
Nominal Group Technique	Identify issues for technical development or marketing of a product, identify key technologies for company (for groups of 8 to 18)	Structure, collects many creative ideas	Structure, does not allow for interaction of ideas	Knowledge of techniques, good group of experts, 4 hours per topic, $5,000 per session
Scenarios	Forecast multiple environments for technologies, strategy analysis	Contingency planning is stressed, forces planning for a changed environment before it occurs	Too qualitative for some, relatively expensive	Knowledge of technique, software is useful, 1 week to 6 months, $4,000 to $40,000
Simulations	Contingency planning, strategy analysis	Forces thought about alternative plans	Requires realistic model which is amenable to sensitivity analysis	Model of topic must already exist, 1 hour to several days, $5,000
Paths and Trees	Disaggregate technology problem in pieces, identify structure, alternative paths	Structure inquiry, identifies planning paths	Normative judgment, ignores alternative goals, rarely predictive	Data which describes all technologies in the dimensions to be graphed, 2 days, $1,000
Portfolio Analysis	Disaggregate technology inventory into sectors for business analysis	Manage existing technology more creatively, profitably	Simply imposing structure may not provide new insights, not predictive of technological success	Data on all company technology which can be categorized, 1 week, $4,000

*Under required resources, the dollar figure is a median estimate of what it would cost in consulting dollars to apply the technique. In other words, this is what a consultant might charge including all necessary labor, data, hardware, and software. Time estimates assume data is either already collected or available electronically.

2   More methods will be hybridized. All three categories of methods examined in this report show strong potential for improvement when used with other methods. Very likely, they can and will grow into product forecasting.

3   The following methods will increase or remain the same in popularity:

—trend extrapolation

—time series estimation

—regression analysis

—historical analogies

—patent trend analysis

—scientific literature analysis

—analysis of user created databases

—interviews

—questionnaires

—idea generation

—nominal group technique

—scenarios

—simulations.

4   The following methods will decline in popularity:

—econometrics

—systems dynamics

—S-curves

—input-output matrices

—Delphi method

—paths and trees

—portfolio analysis.

5   We are aware of only one new forecasting and strategy analysis method that will emerge as a newly popular method in the 1990s, namely the creation of a user's own database for electronic analysis. Obviously, we cannot possibly be aware of all dormant tools of analysis, but we

suspect that some other multi-option tools must be emerging as well because of management's recent recognition that the world is more complex than that modeled by a single forecast. The recent growth in 3-dimensional spreadsheet analysis is evidence of this trend. At any rate, such new tools are highly unlikely to replace the methods discussed above within the next decade.

Recommendations for Managers

Based on these findings, we recommend that managers undertake several actions. In general, we believe managers need to

1   Clarify their needs to analysts so that they can be more responsive.

2   Prepare a written inventory of existing tools and databases presently used by the company. If forecasting and planning are decentralized to different corporate divisions, then consider using a questionnaire. Finally, analyze the inventory to ascertain whether a transfer of existing knowledge, software, or data from one working group to another group would be beneficial.

3   Examine the trends for the 1990s in forecasting tools and explore what these trends mean for the company. Does changing software or data availability suggest any changes the company should make in use of tools? Does an overall increase or decrease in popularity suggest any strategic options for your company? Consider using a group dynamic session with analysts to explore these points.

4   Request analysts to evaluate any specific tools or databases mentioned in this report that promise to improve the company's forecasting ability or provide a better understanding of technology problems.

5   Implement the use of multiple methods in various combinations for better technology forecasting. While there are a great number of possible combinations of methods, one array seems to be particularly attractive:

   —Expert judgment (particularly interviews and surveys) to frame the right question for the forecasting study;

—Expert judgment (particularly idea generation and the nominal group technique, or variations thereof) to identify issues, factors, trends, variables, etc. to be included in the scope of the forecasting study;

—Trend analyses (particularly trend extrapolation, time series, and patent trend analysis) to understand thoroughly the past, present, and most feasible future of each factor in the scope of the forecast;

—Multi-option analysis (particularly scenarios and simulations) to integrate the trends and to generate alternative, including normative, views of the future;

—Expert judgment (particularly idea generation and the nominal group technique) to draw business implications and strategic options from the forecasts;

—Trend analyses (particularly trend extrapolation and time series) combined with other forecasting methods (especially econometrics and financial projections) to do detailed, microscopic analysis for planning purposes.

6   Expand their concept of technology forecasting toward the broader concept of product forecasting, which includes business environment and corporate concerns as well as technological performance.

7   Incrementally integrate changes required by the new technology forecasting methods into the office. This includes acquiring appropriate computer hardware and software, ensuring adequate access to the proliferating number of electronic databases, and training staff to use both new forecasting tools and combinations of tools.

C H A P T E R

6

SOURCES OF INFORMATION

Assessing Technology Strategy. *Technology Forecasts and Technology Surveys, 20* (October 1988), pp 1–2.

Armstrong, J. Scott. *Long-Range Forecasting. From Crystal Ball to Computer.* 2nd ed. New York: John Wiley & Sons, 1985.

Ascher, William. *Forecasting. An Appraisal for Policy-Makers and Planners.* Baltimore: The Johns Hopkins University Press, 1978.

Ashton, W. Bradford, Lawrence O. Levine, and Richard S. Campbell. *Patent Trend Analysis. Tracking Technology Change for Business Planning.* B-TIP Report No. 44. Columbus, OH: Battelle, 1985.

Ashton, Brad and Rajat K. Sen. "Using Patent Information in Technology Business Planning," *Research-Technology Management, 32* (January–February 1989), pp 36–42.

Batabba, Vincent P. "The Market Research Encyclopedia," *Harvard Business Review, 68* (January–February 1990), pp 10–116.

Bjorklund, Glen J. "Planning for Uncertainty at an Electric Utility," *Public Utilities Fortnightly* (October 15, 1987), pp 15–21.

Brody, Herb. "Sorry, Wrong Number," *High Technology Business, 18* (September 1988), pp 24–28.

Campbell, R. S. "Patent Trends as a Technological Forecasting Tool," *World Patent Information, 5* (1983), pp 137–143.

Campbell, R. S. and L. O. Levine. *Technology Indicators Based on Patent Data.* Springfield, VA: National Science Foundation, 1984.

Caton, Christopher N. "Forecast Revision, or What To Do When The Future Is No Longer What It Used To Be." Data Resources of Lexington, Massachusetts, paper presented to the Seventh International Symposium on Forecasting, Boston, Massachusetts, May 1987.

Chu, Kong. *Principles of Econometrics*. Scranton, PA: International Textbook Company, 1968.

Churchill, Gilbert, Jr. *Marketing Research Methodological Foundations*. 4th ed. Chicago: Dryden Press, 1984.

Corn, Joseph J., ed. *Imagining Tomorrow. History, Technology, and the American Future*. Cambridge, MA: MIT Press, 1986.

Delbecq, Andre L., Andrew H. Van de Ven, and David H. Gustafson. *Group Techniques for Program Planning. A Guide to Nominal Group and Delphi Processes*. Glenview, IL: Scott, Foresman and Company, 1975.

Dosi, Giovanni. "Microeconomic Effects of Innovation," *Journal of Economic Literature, 26* (September 1988), pp 1121–1171.

Faruqui, Ahmad. "On the Search for Accuracy in Electric Utility Forecasting," *Journal of Forecasting, 6* (1987), pp 93–95.

Fisher, J. C. and R. H. Pry. "A Simple Substitution Model of Technological Change," *Technology Forecasting and Social Change, 3* (1971), pp 75–88.

Forrester, Jay W. *Industrial Dynamics*. New York: John Wiley & Sons, Inc., 1961.

Forrester, Jay W. *Principles of Systems*. Cambridge, MA: Wright-Allen Press, Inc., 1968.

Forrester, Jay W. *World Dynamics*. Cambridge, MA: Wright-Allen Press, Inc., 1971.

Fowles, Jib., ed. *Handbook of Futures Research*. Westport, CT: Greenwood Press, 1978.

Freudenburg, William R. "Perceived Risk, Real Risk: Social Science and the Art of Probabilistic Risk Assessment," *Science, 242* (7 October 1988), pp 44–49.

Garde, V. D. and R. R. Patel. "Technological Forecasting for Power Generation—A Study Using the Delphi Technique," *Long Range Planning, 18* (August 1985), pp 73–79.

Georgoff, David M. and Robert G. Murdick. "Manager's Guide to Forecasting," *Harvard Business Review, 64* (January–February 1986), pp 110–120.

Geschka, Horst, Ute von Reibnitz, and Kjetil Storvik. *Idea Generation Methods: Creative Solutions to Business and Technical Problems*. Battelle Technical Inputs to Planning/Review No. 5. Columbus, OH: Battelle, 1981.

Goldfarb, David L. and William R. Huss. "Building Scenarios for an Electric Utility," *Long Range Planning, 21* (April 1988), pp 78–85.

Griliches, Z., ed. *R&D, Patents, and Productivity*. Chicago: University of Chicago Press, 1984.

Hawken, Paul, James Oglivy, and Peter Schwartz. *Seven Tomorrows*. New York: Bantam Books, 1982.

Honton, E. J., G. S. Stacey, and S. M. Millett. "Future Scenarios: The BASICS Computational Method." Economics and Policy Analysis Occasional Paper No. 44. Columbus, OH: Battelle, 1985.

Hurter, Arthur P. and Albert H. Rubenstein. "Market Penetration by New Innovations: The Technological Literature," *Technology Forecasting and Social Change, 11* (1978), pp 197–221.

Huss, William R. and Edward J. Honton. "Scenario Planning—What Style Should You Use?," *Long Range Planning, 20* (August 1987), pp 21–29.

Johnston, J. *Econometric Methods*. New York: McGraw-Hill Book Company, 1972.

Kahn, Herman. *Thinking About the Unthinkable.* New York: Avon Books, 1962.

Kahn, Herman and Anthony J. Wiener. *The Year 2000. A Framework for Speculation of the Next Thirty-Three Years.* New York: Macmillan, 1967.

Kahneman, Daniel, Paul Slovic, and Amos Tverskey. *Judgment Under Uncertainty: Heuristics and Biases.* Cambridge, England: Cambridge University Press, 1982.

Kamm, Judith B. "The Portfolio Approach to Divisional Innovation Strategy," *Journal of Business Strategy, 7* (Summer 1986), pp 25–36.

Kinnear, Thomas C. and James R. Taylor. *Marketing Research.* 3rd ed. New York: McGraw-Hill, 1987.

Krogh, Lester C. et al. "How 3M Evaluates its R&D Programs," *Research-Technology Management, 31* (November–December 1988), pp 10–14.

Lee, Thomas H. and Nebojsa Nakicenovic. "Technology Life-Cycles and Business Decisions," *International Journal of Technology Management, 3* (1988), pp 411–426.

Leemhuis, J. P. "Using Scenarios to Develop Strategies," *Long Range Planning, 18* (April 1985), pp 30–37.

Levine, L. O. "Summary of Trends in Photovoltaic Patent Activity," Report PNL-5150. Richland, WA: Pacific Northwest Laboratory, 1984.

Linneman, Robert E. and Harold E. Klein. "The Use of Multiple Scenarios by U.S. Industrial Companies," *Long Range Planning, 12* (February 1979), pp 83–90.

Linneman, Robert E. and Harold E. Klein. "Using Scenarios in Strategic Decision Making," *Business Horizons, 28* (January/February 1985), pp 64–74.

Linstone, Harold A. and Murray Turoff, eds. *The Delphi Method. Techniques and Applications.* Reading, MA: Addison-Wesley Publishing Co., 1975.

Mahmoud, Essam. "Accuracy in Forecasting: A Survey," *Journal of Forecasting, 3* (April–June 1984), pp 139–159.

Makridakis, Spyros and Steven C. Wheelwright, eds. *The Handbook of Forecasting. A Manager's Guide.* New York: John Wiley & Sons, 1982.

Malaska, Pentti. "Multiple Scenario Approach and Strategic Behavior in European Companies," *Strategic Management Journal, 6* (1985), pp 339–355.

Marshall, A. W. "A Program to Improve Analytic Methods Related to Strategic Forces," *Policy Sciences, 15* (1982), pp 47–50.

Martino, Joseph P. *Technological Forecasting for Decision-Making.* 2nd ed. New York: North-Holland, 1983.

McNees, Stephen K. "The Accuracy Keeps Improving," *The New York Times,* (January 10, 1988), p F-2.

Meadows, Donella H. "Charting the Way the World Works," *Technology Review, 88* (February–March 1985), pp 55–63.

Millett, Stephen M. and Fred Randles. "Scenarios for Strategic Business Planning: A Case History for Aerospace and Defense Companies," *Interface, 16* (November–December 1986), pp 64–72.

Millett, Stephen M. *Group Dynamic Methods for Forecasting and Strategic Planning.* Battelle Technical Inputs to Planning/Review No. 23. Columbus, OH: Battelle, 1986.

Millett, Stephen M. "Los Angeles 2007 Scenarios." Economics and Policy Analysis Occasional Paper No. 64. Columbus, OH: Battelle, July 1988.

Millett, Stephen M. "How Scenarios Trigger Strategic Thinking," *Long Range Planning, 21* (October 1988), pp 61–68.

Negoita, Constantin Virgil. *Expert Systems and Fuzzy Systems.* Menlo Park, CA: The Benjamin/Cummings Publishing Co., 1985.

Neter, John and William Wasserman. *Applied Linear Statistical Models.* Homewood, IL: Richard D. Irwin, Inc., 1974.

Parente, Frederick J. and Janet K. Anderson, et al. "An Examination of Factors Contributing to Delphi Accuracy," *Journal of Forecasting, 3* (April–June 1984), pp 173–182.

Pavitt, K. "R and D, Patenting and Innovative Activities," *Research Policy, 11* (1982), pp 33–51.

Pavitt, K. "Patent Statistics as Indicators of Innovative Activities: Possibilities and Problems," *Scientometrics, 7* (1985), pp 77–99.

Phillips, L. D. and L. R. Beach, eds. *Special Issue on Judgmental Forecasting, Journal of Forecasting, 9* (July—September 1990), "editors' remarks."

Porter, Michael E. *Competitive Strategy. Techniques for Analyzing Industries and Competitors.* New York: The Free Press, 1980.

Porter, Michael E. "From Competitive Advantage to Corporate Strategy," *Harvard Business Review, 65* (May–June 1987), pp 43–59.

Quade, E. S. and W. I. Boucher. *Systems Analysis and Policy Planning. Applications in Defense.* New York: American Elsevier Publishing Co., 1968.

Schnaars, Steven P. and Conrad Berenson. "Growth Market Forecasting Revisited: A Look Back at a Look Forward," *California Management Review, 28* (Summer 1986), pp 71–88.

Southern California Edison Company, "Strategies for an Uncertain Future." March 1988.

Stacey, Gary S. *Tech Forecasting and the TECH-RISK-ARRAY: Linking R&D and Strategic Business Decisions.* Battelle Technical Inputs to Planning/Review No. 14. Columbus, OH: Battelle, 1984.

Stacey, Gary S., Joanne C. Hart, Edward J. Honton, Stephen M. Millett, Ingrid Schubert, and Antonio Sfiligoj. *Technology Acquisition Decisions: Using Scenarios for Technology Forecasting.* Battelle Technical Inputs to Planning Report No. 60. Columbus, OH: Battelle, 1988.

Sudman, Seymour and Norman M. Bradburn. *Asking Questions. A Practical Guide to Questionnaire Design.* San Francisco: Jossey-Bass, 1982.

Swager, William L. "Strategic Planning I: The Roles of Technological Forecasting," *Technological Forecasting and Social Change, 4* (1972), pp 85–100.

Swager, William L. "Strategic Planning II: Policy Options," *Technological Forecasting and Social Change, 4* (1972), pp 151–172.

Swager, William L. "Strategic Planning III: Objectives and Program Options," *Technological Forecasting and Social Change, 4* (1972), pp 283–300.

Swager, William L. and Edward S. Lipinsky. "Structuring Ideas for Product Development with Relevance Trees," *American Cosmetics and Perfumery* (May 1972).

Theil, Henri. *Principles of Econometrics.* New York: John Wiley & Sons, Inc., 1971.

Wack, Pierre. "Scenarios: Uncharted Waters Ahead," *Harvard Business Review, 63* (September–October 1985), pp 73–89.

Wack, Pierre. "Scenarios: Shooting the Rapids," *Harvard Business Review, 63* (November–December 1985), pp 139–150.

Zarnowitz, Victor. "The Record and Improvability of Economic Forecasting," National Bureau of Economic Research, Inc., Working Paper No. 2099, December 1986.